Advances in 21st Century Human Settlements

Indexed by SCOPUS

This Series focuses on the entire spectrum of human settlements—from rural to urban, in different regions of the world, with questions such as: What factors cause and guide the process of change in human settlements from rural to urban in character, from hamlets and villages to towns, cities and megacities? Is this process different across time and space, how and why? Is there a future for rural life? Is it possible or not to have industrial development in rural settlements, and how? Why does 'urban shrinkage' occur? Are the rural areas urbanizing or is that urban areas are undergoing 'ruralisation' (in form of underserviced slums)? What are the challenges faced by 'mega urban regions', and how they can be/are being addressed? What drives economic dynamism in human settlements? Is the urban-based economic growth paradigm the only answer to the quest for sustainable development, or is there an urgent need to balance between economic growth on one hand and ecosystem restoration and conservation on the other—for the future sustainability of human habitats? How and what new technology is helping to achieve sustainable development in human settlements? What sort of changes in the current planning, management and governance of human settlements are needed to face the changing environment including the climate and increasing disaster risks? What is the uniqueness of the new 'socio-cultural spaces' that emerge in human settlements, and how they change over time? As rural settlements become urban, are the new 'urban spaces' resulting in the loss of rural life and 'socio-cultural spaces'? What is leading the preservation of rural 'socio-cultural spaces' within the urbanizing world, and how? What is the emerging nature of the rural-urban interface, and what factors influence it? What are the emerging perspectives that help understand the human-environment-culture complex through the study of human settlements and the related ecosystems, and how do they transform our understanding of cultural landscapes and 'waterscapes' in the 21st Century? What else is and/or likely to be new vis-à-vis human settlements—now and in the future? The Series, therefore, welcomes contributions with fresh cognitive perspectives to understand the new and emerging realities of the 21st Century human settlements. Such perspectives will include a multidisciplinary analysis, constituting of the demographic, spatio-economic, environmental, technological, and planning, management and governance lenses.

If you are interested in submitting a proposal for this series, please contact the Series Editor, or the Publishing Editor:
Bharat Dahiya (bharatdahiya@gmail.com) or
Loyola D'Silva (loyola.dsilva@springer.com)

More information about this series at http://www.springer.com/series/13196

Giuseppe T. Cirella

Editor

Sustainable Human–Nature Relations

Environmental Scholarship, Economic Evaluation, Urban Strategies

 Springer

Editor
Giuseppe T. Cirella
Faculty of Economics
University of Gdansk
Sopot, Poland

ISSN 2198-2546   ISSN 2198-2554   (electronic)
Advances in 21st Century Human Settlements
ISBN 978-981-15-3051-7   ISBN 978-981-15-3049-4   (eBook)
https://doi.org/10.1007/978-981-15-3049-4

This Springer imprint is published by the registered company Springer Nature Singapore Pte Ltd.
The registered company address is: 152 Beach Road, #21-01/04 Gateway East, Singapore 189721, Singapore

ai miei genitori (to my parents)

Foreword by Patricia Dale

Sustainability has been a key concept in environmental management since the early 1990s, following the Rio Declaration on Environment and Development of 1992. A major aim of sustainability is to meet current human needs without compromising the welfare of future generations. This has been a difficult concept to implement within many societies and cultures. It involves the complex interactions among all aspects of the environment, with humans both immersed in and dependent on healthy ecosystems. To develop and effectively implement sustainable practices, it demands an interdisciplinary approach which, at least in the early days, was not well understood. There is currently a general and worldwide human settlement pattern of increasing urbanization. Urban development has wide-ranging impacts on natural systems and the ecosystem services they provide. The capacity of humans to alter ecosystems has the potential to damage the systems to and beyond tipping points, from which there may be no return. This is a severe threat to achieving sustainability. To avoid such a consequence requires wide-ranging and interdisciplinary knowledge about ecosystems, their dynamics, how they respond to disturbance, and how best to manage them. This book makes a valuable contribution to this knowledge, with an emphasis on urban systems.

Sustainability is conventionally based on three pillars, at a general level. These are environment, social, and economic dimensions. They are not independent, and this book cleverly brings them together and expands the dimensions with reference to specific case studies. A key strength lies in the interdisciplinary approach, integrating major components with well-argued rationales. The book is organized into three parts. These are sustainability and the environment, economic evaluation, and sustainable strategies in urbanization.

The first part focuses on sustainability and the environment. It takes an in-depth view of human–nature relationships from a global perspective and with reference to historical and economic factors. It stresses the importance of choice in making decisions about the environment. The five chapters in this section explore specific concepts including carrying capacity, human migration, and associated land use issues. Case studies illustrate the points raised. It concludes with an integration of scales about how the human species shares the planet with other species.

The second part considers economic issues. The three chapters explore how people deal with economic aspects of production, consumption, money, and resource management. It explores economic evaluation and uses a mathematical model in the context of land speculation which underlies processes of urbanization. Issues of peri-urban development are addressed with case studies exemplifying the points raised. It considers reasons for urbanization and concludes by emphasizing the potential role of economic evaluation for achieving urban balance and sustainability outside mainstream economics.

The third part links urbanization and sustainability. The three chapters in this part show that populations are becoming increasingly urban and that this is a challenge for urban planning. It advocates the need for ecosystem services to be incorporated, via ecological knowledge, into urban design. Incorporating green-friendly areas and appropriate infrastructure will result in resilient and sustainable urban systems, providing humans with high levels of urban amenity while not compromising the provision of ecosystem services. This part integrates the three pillars of sustainability (i.e., environment, social, and economics) culminating with a case study illustrating impacts of urbanization including constraints on land use. The authors are to be commended on writing a thoughtful book that will promote informed discussion about urban sustainability based on convincing case study support. This will increase our overall understanding of sustainability, especially in a rapidly urbanizing world.

Patricia Dale
Emeritus Professor
School of Environment and Science
Environmental Futures Research Institute
Griffith University
Brisbane, Australia

Foreword by Monika Bąk

The slogans related to sustainable development are nowadays one of the most frequently appearing in social spaces. We can hear and read about sustainability in the media, scientific publications, and political debate. Scientists, politicians, social activists, as well as governmental, non-governmental, and even religious organizations are involved in the debate. People may even feel overwhelmed and ask themselves what the argument is actually about, i.e., whether it is relevant and whether it really affects our lives and communities. Indeed, in this global discussion we are witnessing extreme views that include the most expressive and striking—on the one hand climate change hysteria and, on the other, an underestimation and disregard for danger—undermining rational argumentation. Thus, observation of this debate on sustainable development may also cause some confusion due to the interdisciplinarity that increasingly includes more and more areas of life and involves a diversification of subject matters. At the forefront, the argument begins with ecological, social, and economic spheres.

At present, there is deep discussion that integrates technological development, sociological and psychological factors, ethics and philosophy, as well as many other multi-disciplinary fields. Moreover, new linkages and sub-disciplines have been developed in relation to sustainability, e.g., sustainable economic development, ethical foundations of sustainability, philosophical perspectives on sustainable development, sustainability and law, and human ecology—to mention a few. As a result, internal contradictions primarily between economic and ecological aspects of sustainable development play an important part in both macro- and micro-level advancements. Obviously, economic profit, as it historically has been established, is in contradiction to a healthy environment (i.e., specifically via the monetization of everything) in which profit-driven businesses believe they are the means to an end. Hence, as the microeconomic approach changes, considerable importance will need to be focused on corporate social responsibility and the like.

Many years were required in order to reach a general consensus that the progress of civilization has resulted in the devastation of the natural and social environment. This is the opinion with which at least the majority (i.e., not all) agree. Costs of improvement, waste, lack of moderation, wealthy societies, social stratification, and consumptionism are all features from the last few decades which have negative cause-and-effect environmental risks. Much still needs to be done to raise awareness that in our global village we should protect it for future generations. Still, there are some typically egoistic concepts and approaches by actors (e.g., forms of individualism, lobbying groups, and even some organizations and governments) promoting here-and-now (i.e., business-as-usual) ideas of happiness. This is not, in any case, a new problem, but is reflective of specific psycho-structural thinking of human beings. There is a saying in Polish "*po nas, choćby potop*" which means "after us, even a flood" which supposedly originated from France (i.e., "*après moi, le déluge*") during the century and history of King Louis XV. According to possible interpretations, the phrase may have been coined by Madame de Pompadour, the lover of the king, who in the opinion of the people wasted state money on the pompous games at Versailles and urged the king to be particularly extravagant— referring to after the king's reign (i.e., death) the nation will be ruined and hence caring for the future as meaningless. Similarly, Poles currently use the phrase when assessing the view that nothing that happens in the future matters. Beyond the philosophy, it would seem that this is where sustainability and future generational thinking must begin. As such, what is needed most is a thorough analysis of the problems and an attempt to solve them. Knowledge base and know-how to this dilemma can be found throughout this book.

Editor, Prof. Dr. Giuseppe T. Cirella, for years has been engaged in scientific research and popularization-based activities in the field of human and nature relations. A feature of his background is his pragmatism in practical applications, skill in facing current scientific challenges, and ability to seek answers and propose action that regularly promises effective results. Professor Cirella partakes within several scientific disciplines, joint research within various interdisciplinary fields, and international teams. This book is a result of the editor's activity in recent years. Together with the Polo Centre of Sustainability in which he is the founder, he organized the International Conference on Sustainability, Human Geography and Environment 2018 in Cracow, Poland, which attracted delegates from all over the world. He is regularly invited to keynote at international conferences, e.g., twice in the last year to China, where he presented a number of co-relating issues on environmental threats within the country. In 2018, he partook in a professorship at the University of Pretoria, Pretoria, South Africa, in which he examined problems of inequality and multiple southern African country imbalances. Professor Cirella is currently employed at the Faculty of Economics, University of Gdansk, Sopot, Poland, yet another place, developing from post-communism in which macroeconomic outcomes are significantly lower than Western economies. As a backdrop to his extensive cross-cultural expertise, Prof. Cirella continues to directly observe societal and sustainable development changes from a worldly view. The advantage of the approach presented in this book is its versatility, as well as the silhouette of the editor, who has direct insight into what is happening in practice and what is being discussed from his international experience. The pragmatic approach and the invitation of authors from all over the world are case study-specific. The book is divided into three sensible and logically thought-out parts. In Part One, we find an order to the concepts, indication of how to understand the relationship between human beings and nature, and reflection on the contemporary condition. In Part Two, economic characteristics explore important inflammatory points and source the conflict between the goal of profit and threat. In Part Three, attention is focused on the city using relating urban-based experimentation. Urban strategies play an important role in sustainability since urban surroundings make up the largest area where people live—forming key concerns for the human–nature relation lens. I invite the readers to relish this interesting and diverse book in which the authors from many corners of the world shared their reflections and research results.

Monika Bąk
Dean
Faculty of Economics
University of Gdansk
Sopot, Poland

Foreword by Yi Xie

This is a book with significant importance regarding growing concerns between human–nature relations in the context of ever-increasing levels of urbanization—worldwide. It gives me great pleasure to support Prof. Dr. Giuseppe T. Cirella's invitation to write this Foreword and provide useful hints in better understanding the scope of this book. Sustainability is one of the most critical and sensible concepts when considering human–nature relations. We should remember that human society originated from (and is developed by) nature itself and civilization is sustained by a multiplicity of natural resources and environmental services which nature mostly provides for free. Nature as a resilient, adaptive, renewable system evolves in a sustainable manner when human disturbance does not exceed its carrying capacity. This can be referred to as a human–nature balance. However, this balance easily is broken when unconstrained development, especially within fragile natural systems, exists. An imbalance in human–nature relations makes both human society and the natural system worse-off—equating to increased levels of unsustainability. If non-sustainable development is prolonged, in any form, a re-understanding from a societal level should be carefully considered. This is the primary motivating factor in the first part (i.e., of three) of the book.

The challenges of sustainability can also be measured by economic means. Concepts such as scarcity of natural resources determined jointly by supply (i.e., from the nature system) and demand (i.e., from human society) are examined. Given supply from the natural system, increasingly larger demands have been causing severe levels of scarcity, forcing natural resources to become ever increasingly valued and profitable. In the context of effective monitoring and sanctioning of resource extraction, scarcity (i.e., together with high profit) drives a highly motivated human system to take more than the natural supply can provide (i.e., human beings are exceeding natural levels of resource rejuvenation). As a consequence, it can be said that the natural system is dysfunctional from a human viewpoint and cannot properly maintain a sustainable, operative supply for the long term. Land scarcity issues, as a result, are discussed in parallel to rapid urbanization concerns. The second part provides a unique economic perspective that considers some of these issues in relation to the human–nature interface by focusing on control factors of how human beings maneuver within the economic pillar of sustainability.

Urbanization bears the fruit of human ingenuity. Booming urban development is expected to provide more employment, enhance allocation and efficiency of natural resources, and promote human welfare. These transformative effects will, however, trigger a human–nature response (i.e., due to the sizeable use and speeding up of natural resource consumption from the built environment). Urbanization, overall, dominates much of the human experience. As such, a large amount of people living in more compact urban environments will accumulate additional pollution levels in which the natural system will need to offset. Urbanization, still heavily embraced by lesser developed countries, requires our full attention. Key queries that integrate how to incorporate urbanization into sustainability and the human–nature relation paradigm are brought to the forefront in the third part.

The complexity of the human–nature relationship is the founding core of this book. Understanding this relationship is a challenge many scholars and policy makers alike have examined in terms of sustainable development. This book simplifies some of these complexities by adopting case studies and profiling a number of country-specific examples. The advantage of case study research is obvious; the case studies include long-term observation and original scenario-based experimentation. The case studies are well supported in terms of argument, viewpoint, and reflectivity of the real world—in terms of the Global North and Global South. I sincerely feel this book will enhance our understanding of sustainable development in the context of urbanization by adopting the three perspectives laid out in conjunction with the analytical instrumentation presented. In all, the book contributes to better solutions for sustainable urbanization in an era where technological advancement interlinks humanity like never before.

Beijing, China   Yi Xie
                                                Professor of Forest Economics
                                        School of Economics and Management
                                                Beijing Forestry University

                                                        Fulbright Scholar
                                        Center for Conservation Biology
                                                University of Washington

Foreword by Sathish Kokkula

It is my honor to have received an invitation to write this Foreword for my continental colleague, Prof. Dr. Giuseppe T. Cirella, where I believe deeply in the educative value of interpretive discussion, among many, within the multi-discipline of sustainability. As a first perspective, this book examines a range of specific issues from the Global North to Global South in which interwoven concepts of human settlement and differentiation are looked at. As human–nature relations are a balance between the philosophical and scientific, key messages from the book are broken down as follows: sustainability issues, economic evaluation, and urbanization strategies. Moreover, the book highlights the importance of interdisciplinary perspectives for a deeper understanding of the human–nature interface. The core essence of the book examines the integral relationship of human society and nature.

The first part emphasizes sustainability thinking in terms of a global perspective with insight from historic as well as economic viewpoints. A case study extrapolates the sociocultural carrying capacity where impacts of population growth are considered in terms of increased population of residents and floating tourists. Evidence to safeguard local identity, heritage, and social coexistence are assessed. Next, territorial integration of multicultural environments, i.e., the mixed couple phenomena emerging within Italian society, is reviewed. Further, sustainable land reforms are shown to improve migration management, in particular, discussed from an African context exemplifying historical land injustice, implementation of land reforms, and mitigating marginalization. Specific land reform legislation within the EU-Africa partnership on migration management is presented as an example of intercontinental cooperation for human security. In terms of nature conservation, the Western Balkans is highlighted with historical emphasis placed directly upon political conflict and the relationship it had on contemporary nature-protected areas.

In the second part, the focus shifts on how human beings attentively employ economic efficiency via varying mathematical formulations—explored in terms of urban land speculation practice, contemporary urbanization, and settlement patterns. In particular, a case study into investment and development modeling methodologies is looked at from an administrative level. The scope includes economic evaluation and specifically examines an economic umbrella of concepts outside the mainstream of economics. Cohesion policy for Europe 2020 is also reviewed by showing harmonization development discrepancies to decrease levels of disparity. Cohesion policy is focalized on the European Union Member States and region. At length, a multiple of concepts interlinking urbanization and sustainability are assessed where statistical data from Eurostat and the United Nations are reviewed. Urban growth, together with the urban landscape, link development, livelihood, and the interlinking rapport between human–nature relations are also considered. In this context, techniques of geographical information systems together with unmanned aerial systems highlight another case study that emphasize best urban planning (i.e., by way of differentiating flood maps).

In the third part, a focus on improving quality of urban life and human well-being, sustainability concepts of ecosystem services, and urban biodiversity is clearly described. The need for sustainable urban strategies that integrate ecology in city design by way of environmental, social, and economic action is elaborated. Urban land use and population density change evaluate potential cause-and-effect phenomena in regard to the projected urbanization boom. Results from the varying case studies exemplify excellent experimentation and decades of hard work and experience. Reading this book, you will find it hard to defend the view that good authors are born, not made.

Bengaluru, India   Sathish Kokkula
                                                Research Engineer and Project Consultant
                                                       Polo Centre of Sustainability
                                                                   Imperia, Italy

                                                            Inspection Team Leader
                                                Organization for the Prohibition of
                                                              Chemical Weapons
                                                   The Hague, The Netherlands

Foreword by Benny Mantin

Our world is rapidly changing, human populations are increasing, globalization is impacting trade patterns, progress in automation increases our productivity, and medical innovations improve our health. All in all, our quality of life and welfare are improving. However, many of the aspects of our prosperity come at a price as they carry substantial implications to the climate and natural environments. These man-made changes lead, for example, to decimation of forests, reduction and fragmentation of natural areas, deep scars due to mining, desertification, and extinction of many species. These ultimately result in migration and conflicts over ever-dwindling resources.

A major impact on a global scale is the changing pattern of carbon inventories. There is a constant flow of carbon between the atmosphere, the oceans, the crust (namely fossil fuels), and terrestrial ecosystems. In 2016, the natural carbon flow resulted in a net removal of 5 GtC of carbon inventories from the atmosphere into Earth every year. However, due to human emissions, which released 11 GtC of carbon into the atmosphere, the atmosphere's inventory carbon increased by a total of 6 GtC. To cap the warming level at 2 °C by 2100, according to science, we shall limit the atmosphere's carbon inventory at 1318 GtC implying human activity needs to remain within 10.6 GtC per year. If we wish to maintain the current inventory levels, we actually need to halve our emissions. This requires us to dramatically change our behavior and how we treat the nature and the environment.

There are already examples of successful transitions. Consider, for instance, global supply chains, which inherently suffer from limited level of transparency thereby facilitating fiercer competition as one progresses up the supply chain closer to raw material, resulting in, for example, conflict minerals and labor abuse in less developed countries. It is estimated that the environmental costs externalized from global production systems amount to $4.7t per year. As incentives are not aligned and given the lack of traceability to the source, such outcomes whereby environmental costs are externalized emerge. Recently, several companies started changing their sourcing practices, oftentimes due to external pressure. Most noticeable is the change going through the apparel industry. The mounting pressure is shifting this industry, which heavily relies on deep supply chains consisting of 6 tiers or more (i.e., keeping in mind that generally most companies do not have visibility beyond their first-tier suppliers). This industry now starts embracing more sustainable practices. Such initiatives include the Higg Index developed by the Sustainable Apparel Coalition, the environmental profit and loss account (e.g., Puma), footprint chronicles (e.g., Patagonia), and many others. Clearly, these examples provide us with reasons for optimism.

To conclude, we are heading to a crossing point that will require us to make some tough decisions, which will require us to change many of our fundamental behaviors, which will induce us to think in a more holistic manner and force us to develop new methods and embrace a rapid wave of innovation. As our global population is becoming ever more literate, increasingly awareness of the consequences and the importance of human–nature relations, I trust we will find the right path forward.

To change our behavior, we first need to assess and better understand the human–nature relationship. This book contributes to an important and topical body of research that expands our knowledge on this matter by considering sustainability and the environment, with regard to economic perspectives of such interfaces, and focusing on urbanization while accounting for sustainability. Such valuable studies can feed into the decision process and inform policy makers and regulators about available solutions. To that end, this book provides some of the fundamentals that shall pave the way.

Benny Mantin
Professor
Faculty of Law, Economics and Finance
University of Luxembourg
Luxembourg City, Luxembourg

Preface

In reference to the development of sustainable societies, there is a critical scope in terms of human interconnectedness with the world around us and the noise society bares. Noise, in this sense, is the busyness societies, especially contemporary, levy on an individual. If one were to assess this levy, it could be labeled, respectively, as weight. In a sense, it would be an individual's level of effectiveness or aptitude to participation within society versus one's unproductiveness or imaginative state of thinking "outside of the box." Societies, especially contemporary, face diverse challenges that need to acknowledge functional, versus dysfunctional, action. This acknowledgment, evident from reviewing the chronology of art and usage of modern-day social media, relates to a growing worldwide concern of ideas and concepts that people from all scopes of life are probing. This concern correlates the human necessity of need and want, at the individual level, and its coexistence and framing via day-to-day living. The level of harmonization societies exert is somewhat of a balancing act in which large scoped challenges—such as rising inequality, loss of biodiversity, and armed conflict—are at the core of bandage-like fixes that have been relatively inept. The need to rearrange human–nature relations is fundamental to trying to comprehend the noise in which functionality, between society and nature, defines human sustainability. Sustainable human–nature relations merge key interdisciplinary fields of environmental management, sociology, human geography, urban development, economics, ethics, and philosophy. This book is a compilation of work interlinking a number of these fields bridging its three parts: environmental scholarship, economic evaluation, and urban strategies.

Sopot, Poland

Giuseppe T. Cirella

Contents

Editor and Contributors

About the Editor

 **Prof. Dr. Giuseppe T. Cirella** Professor of Human Geography, works at the Faculty of Economics, University of Gdansk, Sopot, Poland. He specializes in development and environmental social science, human geography, and sustainability. His interdisciplinary background also includes socio-political research throughout Eastern Europe, Africa, and China. After completing his doctorate (Ph. D.) at Griffith University, Australia with-in the Centre for Infrastructure Engineering and Management, entailing the development of a sustainability-based index, he founded the Polo Centre of Sustainability (www.polocentre.org) in Italy. Notably, he has held professorships and scientific positions at Saint Petersburg State University, Saint Petersburg (Russia), Inner Mongolia University of Science and Technology, Baotou (China), Life University, Sihanoukville (Cambodia), and Free University of Bozen, Bozen (Italy). In his early career, he worked with the Canadian International Development Agency in Indonesia as well as with RADARSAT International in Brazil.

Contributors

Birhanu G. Abebe Ethiopian Institute of Architecture, Building Construction and City Development, Addis Ababa University, Addis Ababa, Ethiopia

Solomon T. Abebe Polo Centre of Sustainability, Imperia, Italy

Piroska Ángel Observatory of Cities UC, Pontifical Catholic University of Chile, Santiago, Chile

Federico Benassi Italian National Institute of Statistics, Rome, Italy

Kay Bergamini Institute of Urban and Territorial Studies, Pontifical Catholic University of Chile, Santiago, Chile

Sutatip Chavanavesskul Department of Geography, Srinakharinwirot University, Bangkok, Thailand

Shanshan Chu School of Geography, Nanjing Normal University, Nanjing, China

Giuseppe T. Cirella Faculty of Economics, University of Gdansk, Sopot, Poland

Bedane Sh. Gemeda Ethiopian Institute of Architecture, Building Construction and City Development, Addis Ababa University, Addis Ababa, Ethiopia

Tianwu Ma School of Geography, Nanjing Normal University, Nanjing, China

Samuel W. Mwangi Institute of Political Science, Tübingen University, Tübingen, Germany

Alessia Naccarato Department of Economics, Roma Tre University, Rome, Italy

Andrzej Paczoski Faculty of Economics, University of Gdansk, Sopot, Poland

Matjaž N. Perc Faculty of Civil Engineering, Transportation Engineering and Architecture, University of Maribor, Maribor, Slovenia

Tea Požar Institute of Geography, University of Bamberg, Bamberg, Germany

Alessio Russo School of Arts, University of Gloucestershire, Cheltenham, UK

Raden G. Shaumirahman Urban and Regional Engineering Faculty, Pasundan University, Bandung, Indonesia

Agus Supriyadi School of Geography, Nanjing Normal University, Nanjing, China

Tao Wang School of Geography, Nanjing Normal University, Nanjing, China

Abbreviations

ASCIs	Areas of Special Conservation Interest
CA-Markov	Cellular automata-Markov
CBD	Central business district
DEM	Digital elevation model
DSM	Digital surface model
ECFR	European Council on Foreign Relations
EU	European Union
FGDs	Focus group discussions
GCPs	Ground control points
GIS	Geographic Information System
ICT	Information and communications technology
IPBES	Intergovernmental Science-Policy Platform on Biodiversity and Ecosystem Services
Istat	Italian National Institute of Statistics
IUCN	International Union for Conservation of Nature
LCM	Land Change Modeler
MAB	International Coordinating Council of the Man and the Biosphere
MLP	Multilayer perceptron
MRC	Mombasa Republican Council
MS	Member State
NASA	United States National Aeronautics and Space Administration
RCE	Regional competitiveness and employment
RGB	Red-green-blue
RTD	Research and technological development
SDGs	United Nations Sustainable Development Goals
SEZ	Special economic zone
SfM	Structure from motion
SLDF	Sabaot Land Defense Force

TJRC	Truth, Justice and Reconciliation Commission of Kenya
UAS	Unmanned aerial system
UAV	Unmanned aerial vehicle
UNESCO	United Nations Educational, Scientific and Cultural Organization

Sustainability: Understanding and Insight

Human-Nature Relations: The Unwanted Filibuster

Giuseppe T. Cirella, Samuel W. Mwangi, Andrzej Paczoski, and Solomon T. Abebe

Abstract Human–nature relations encompass many of the age-old questions about our existence, place, and time. This chapter explores some of these notions and offers insight into the question "why protect nature?", the Gaia theory, and linkages from a historical and economical viewpoint between the Global North and the Global South. Arguments in regard to moral and utilitarian viewpoints explore nature conservation with respect to ecocentrism versus anthropocentrism. Gaian ideology is defined and used as a premise to tie sustainability and human responsibility to human–nature and human–human relations. Example research interplays between the Global North and the Global South as two subsystems of human settlement. We utilize Africa as an example of the Global South subsystem and the global economy as an indicator for differentiation. In addition, the objectives, i.e., a recap, of the book and synopsis of the individual chapters are presented.

Keywords Human settlement · Habitation · Gaia theory · Global North · Global South

1 Key Message, Objectives, and Organization of the Book

To different people, the idea of human–nature relations may pose a dilemma, lead to questions much of what this book encompasses, and stimulate useful debate. The

G. T. Cirella (✉) · A. Paczoski
Faculty of Economics, University of Gdansk, Sopot, Poland
e-mail: gt.cirella@ug.edu.pl

A. Paczoski
e-mail: andrzej.paczoski@ug.edu.pl

S. W. Mwangi
Institute of Political Science, Tübingen University, Tübingen, Germany
e-mail: sawaamy@gmail.com

S. T. Abebe
Polo Centre of Sustainability, Imperia, Italy
e-mail: solomtu6@gmail.com

© Springer Nature Singapore Pte Ltd. 2020
G. T. Cirella (ed.), *Sustainable Human–Nature Relations*,
Advances in 21st Century Human Settlements,
https://doi.org/10.1007/978-981-15-3049-4_1

unwanted filibuster and lengthy debate—obstructive thoughts, notions, and actions—yearn for change. Human beings and their expansiveness to settle Earth, and in that expansiveness, nature—our home—is the interplay human beings, in that spirit, innovate and develop settlements. Human–nature relations is a balancing act of ethos and science. The debate concerning this relationship is multidisciplinary with a vast array of issues and perspectives. This book examines three signature components of the relationship: (1) sustainability and the environment, (2) economic evaluation, and (3) urbanization in conjunction with sustainability-oriented strategies. The book argues the use of interdisciplinary perspectives to facilitate a deeper understanding of the complexities of each component at the human–nature interface. The core argument is survival and our direct connection with the natural environment. Quoting Schultz [1], "[…] achieving a sustainable lifestyle depends on establishing a balance between the consumption of individuals, and the capacity of the natural environment for renewal." Any separation from nature—"as if we can get along without it" [1]—is only temporary at best and arguably not even that. Earth's wonder can leave us in awe with beauty while still leaving us at the mercy of backlash (i.e., by natural disasters). All forms of human settlement (i.e., infrastructure, agriculture, transport, and any other aspects of the built environment) can segregate human beings from the natural environment itself. This book examines the inherent connection human society makes between humanity (i.e., self) and nature, and the impact of built environments on these implicit cognitions.

The organization of the book is divided into three parts. Part One begins with an in-depth look into sustainability thinking by developing a better understanding of our role on Earth as well as awareness from a global perspective (i.e., this chapter and Chaps. 2–5). The greater questions of human–nature relations are taken into consideration with insight from a historic and economic viewpoint as well as the divide between the Global North and the Global South. This chapter sets the stage for the book with emphasis on capturing the reader's interest and elucidating on the enormity and importance of human choice and the subsequent issue of managing such decision-making. Case studies into sustainable population in conjunction to carrying capacity (i.e., Chap. 2) as well as social sustainability research into territorial integration of foreigners (i.e., Chap. 3) are investigated. As a precursor to social-oriented findings, comprehensive irregular migration is explored by looking at land-related issues that can be associated with direct and indirect causes of instability, conflict, and violence (i.e., Chap. 4). A nature-based study on the ecological network, Emerald, concludes with an international rendering on the importance of controlling, preserving, and protecting nature (i.e., Chap. 5). Sustainability frames these chapters by piecing together a continuum from the greater and outer scope of Earth and our role as a species, to how we choose to live among fellow human beings and the species we share this world with.

Part Two shifts the attention to how human beings generally cope with efficiency in terms of production and consumption, supply of money, and management of available resources—the economy. Economic evaluation is explored in three subsequent chapters (i.e., Chaps. 6–8). Exploratory research examines urban land speculation, via mathematical investigation, by piecing together challenges of contemporary urbanization and settlement patterns that influence principal actors. Specifically, economic forces driving peri-urbanization, land markets, and informal stimuli are explored. This research encompasses an attempt to discover the process of land speculation in peri-urban and urban land by looking at the principal reasons land speculation occurs (i.e., Chap. 6). Next, a case study into investment and development modeling methodologies, examined at the administrative-level, monitors impact on society, environment, security, and public health (i.e., Chap. 7). Lastly, a comprehensive look at cohesion policy presents elements for human integration as well as mechanisms to develop policy convergence via the elimination of development inequalities (i.e., Chap. 8). The scope of including economic evaluation, and specifically these three chapters, is to collate an economic umbrella of concepts that highlight the potential for balance and cooperation outside the mainstream of economic thinking.

In Part Three, multiple concepts interlinking urbanization and sustainability are investigated (i.e., Chaps. 9–11). United Nations [2] statistics indicate that 55% of the world's population live in urban areas—this is expected to increase to 68% by 2050. Urban growth together with the urban landscape will be a challenge for development, livelihood, and human–nature relations—*per se*. Best urban planning practices will include developing urban green-friendly areas and infrastructure capable of coping with environmental monitoring of related natural and urban systems (i.e., Chap. 9). Exploratory sustainability concepts will depend on ecosystem services and biodiversity to improve quality of urban life and human well-being. Sustainable urban strategies will need to integrate ecology in city design and support resilience-oriented development via ecological, social, and economic action (i.e., Chap. 10). Finally, a case study on potential cause-and-effect phenomena of urban land use and population density change shows underlying constraints, by using the example of an urban corridor, to the projected urbanization boom (i.e., Chap. 11).

This book utilizes case studies and examples from around the world and collates a compendium of scientific research and intelligentsia with universal relevance. An examination of philosophical thought interplays with humanity's aptitude to live among fellow human beings and the species (i.e., nature) we share and co-depend on for co-survival. Expansive studies look at state-of-the-art research in economic evaluation and urban-oriented strategies as relational aspects of the bigger picture of why and how protecting nature is important as well as the inner compass of morality, virtue, and righteousness to the individual, natural world, and beyond.

2 Why Protect Nature?

Human beings, as a "great force of nature", have been altering ecological and evolutionary processes across the Earth for millennia [3–6]; however, in the last half century unprecedently levels of global change to land and oceans show signs of an outward imprint in each of its corners [7]. While the question "why protect nature?" conveys a very basic query philosophers, ecologists, and conservationists alike have been debating, much before the Industrial Revolution, it is clear the perspective from after this period dwarfs the imaginative view of our predecessors. This idea engages not only the capacity but the generational decisions on what mass-Earth-altering action will take place. This action, specifically, refers to mass land alteration from continued agriculture, industry, and urbanization. With this in mind, a modern philosophical viewpoint on what persuades human beings to preserve and conserve species confronts our expansiveness to develop the Earth with or without them. This dilemma is between human settlement and habitation versus all other species' habitable space to live (i.e., in reference to an existence in one's natural surroundings). Two such arguments, in this regard, are discussed: (1) moral (i.e., other species having intrinsic value) and (2) utilitarian (i.e., other species having economic value to humans). In the context of human activity and the environment (i.e., the arguments to the title of this section), it serves as a basis to guiding how we behave with respect to protecting nature. An overview of these arguments is explored with respect to ecocentrism versus anthropocentrism.

To begin, the moral argument conservationists have long held is that nature should be protected simply because it has valuable in itself—this is based on collective human history of moral thought. Soulé [8] argued that human beings should safeguard endangered ecosystems due to the fact that they ethically have intrinsic value that motivates conservation [9, 10] and that extinction is immoral [11]. However, more recently, some ecologists have criticized this line of reasoning. For example, Kareiva and Marvier [12] contend that ethical arguments alone have not sufficiently prevented the rise of this problem—such as large-scale biodiversity loss (i.e., species extinction), deforestation, habitat loss, and augmented levels of pollution.

Over the last five millennia, human activity has brought about major changes in ecosystems worldwide [13], as such we have entered what scientists refer to as the Holocene extinction, otherwise referred to as the sixth mass extinction, as a result of human activity. This large level of extinction spans numerous families of plants and animals, including mammals, birds, amphibians, reptiles, and arthropods [14]. A key source of this extinction is the expansion of human settlement and the creation of crop fields for agriculture and livestock. The persistent clearing of forests, wetlands, and other wild areas have been steadily declining habitats of almost every kind of plant and animal—globally [13]. From this standpoint, the conservationist argues that a moral dilemma exists and that the critical challenge is between self-interest and the other. Accordingly, conservation ethics argues that a viable, healthy natural environment is consequential to a viable, healthy human being. Thus, it can be said that self-interest on Earth represents a symbiotic relationship within the scope of overlapping systems of ecology and survival.

The utilitarian argument, on the other hand, dates back to David Hume (1711–1776) and Jeremy Bentham (1748–1832) and the works of John Mill (1806–1873), in particular, Mill's writing from his book "Utilitarianism" in 1863 [15]. Mill's consequentialist theory of moral philosophy is solely built on the notion that consequences or outcomes of an action determine moral worth. The actions which increase welfare can be considered as correct and contrariwise incorrect through monetary measuring of utility change. The two most widely used techniques are the indirect approach (i.e., travel cost method) and direct approach (i.e., contingent valuation method). Recently, Kleijn et al. [16] published, in Nature Communications, how useful (i.e., persuasive) economic arguments are for conserving wild bee populations from around the world. They reviewed 90 studies on wild bees spanning five continents and about 1400 agricultural fields. They found that wild bees are valuable for crop pollination contributing over US $3000 per hectare to the production of insect-pollinated crops. In other words, the focus on conserving nature is because of its usefulness (i.e., via ecosystem services) to human beings [12]. By focusing on ecosystem services, nature provides free-of-cost services which have value (e.g., natural pollination of crops, filtration of clean drinking water, and decomposition of waste). By putting a monetary value on these services, a more powerful case for conservation or protection can be made. However, looking back at Kleijn et al.'s [16] wild bees pollination research, it was found that only a small percentage of species were doing most of the pollinating (i.e., about 12% where the studies took place and only 2% of the species in the regional species pool). This suggests that a small number of species dominate the contribution of wild bees to crop production value. This indicates that nature has inherent value and that wild bees that do not pollinate crops may still provide valuable services—or at least a supportive role—in the overall make-up of a healthy ecosystem. Namely, wild bees that do not pollinate crops can play an important role in the life cycle of other plants (i.e., including flowers) as well as produce honey, wax products, and anti-bacterial components; thus, it would be an unreliable assumption to assume these wild bees have no value. Rachel Winfree, an ecologist at Rutgers University and co-author of the Kleijn et al. [16] study, stated "[…] if you start losing a lot of plant species, ecosystems can start collapsing" [17]. Kleijn et al. [16] concluded that for species that do not have obvious economic value, ecologists and conservationists will have to rely on moral appeals—as discussed previously. Nonetheless, putting monetary value on ecosystem services can be seen as somewhat controversial or misconceived by non-economists even though most economists accept this practice with varying levels of discrepancy over how it should be enacted [15].

As an offshoot to the dilemma between moral and utilitarian arguments, Chan et al. [18] wrote that instrumental and intrinsic values may fail to resonate with views on personal and collective well-being, or "what is right," in regards to nature and the environment (Fig. 1). Without a complementary outlook on the way value is expressed, and understood, such a focus may inadvertently promote a worldview at odds with a fair and desirable future. The concept of relational values in regard to human–nature relations considers ethical principles that can promote environmental stewardship, conjugated with recognition of nature's contributions, for human benefit. Relational values are defined as preferences, principles, and virtues associated with relationships, both interpersonal and jointed by policies and social norms [18]. In the environmental sphere, these are connected with nature's contributions to people and culturally specific understandings of what "leading a good life" means, as defined by the Intergovernmental Science-Policy Platform on Biodiversity and Ecosystem Services (IPBES) framework [19]. Hence, this has given chance to

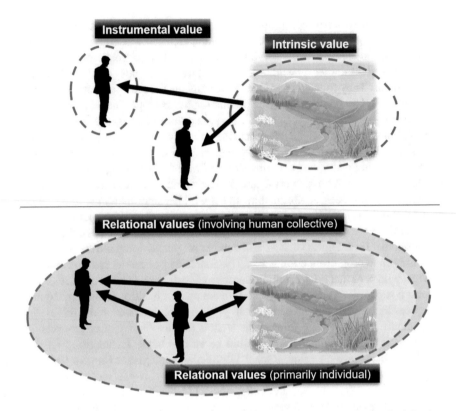

Fig. 1 Instrumental, intrinsic, and relational values [15] where instrumental value is defined as being in and seeing nature brings people pleasure or satisfaction; intrinsic value is defined as nature having value, independent of people; relational value (involving human collective) is defined as four points: place is important to people within my community and to who we are as a people (i.e., cultural identity); being in nature provides a vehicle for an individual to connect with people (i.e., social cohesion); caring for ecosystems is crucial for all fellow humans, present, and future (i.e., social responsibility); and caring for all life forms and physical forms is a moral necessity (i.e., moral responsibility to non-humans); relational value (primarily individual) is defined as three points: place is important to the individual "to who the individual is a person" (i.e., individual identity); personal care for land fulfills the individual and helps the individual lead a good life (i.e., stewardship eudaimonic); and keeping the land healthy is the right thing to do (i.e., stewardship principle and virtue)

reframing of the discussion and debate about environmental protection and, potentially, more productive policy approaches (e.g., Piccolo's [20] line of reasoning for nature conservation).

Generally, recognizing human relational values (i.e., either involving human collective or primarily individual), in addition to instrumental and intrinsic value, may indeed be considered another creative approach to solving the difficult and short-term tradeoffs concerning nature conservation and human well-being. Solving environmental problems and newly emerging issues that threaten nature demands new economic models [21], broader philosophical and scientific thinking [22, 23], and ecocentric ways of teaching and learning [24]. This can relate to the metaphor that we are all passengers bound on a ship, with no way off, all destined to the integrity and rules of the ship, and if it breaks apart, regardless of class, everyone perishes—hence it is in the interest of all to come to common accord. This metaphor represents part of the entanglement of human–human responsibility and relations and, in many ways, parallels the greater scope of human–nature relations.

3   One Ship, Two Classes

To paraphrase Lovelock [25], planet Earth was meant to be a living organism float-ing in the space. Lovelock's theoretical framework is known as the Gaia theory. The Gaia theory shapes our understanding of co-evolvement of the Earth, the environ-ment, life, and humanity. One ship, two classes expounds the relevance of the Gaia theory in understanding the Earth, environment, and human habitation as well as the developed versus developing world (i.e., two classes). We argue that the Earth and the environment are intertwined and unless the international community embraces this co-constituent relationship, sustainability will not be achieved. Secondly, look-ing at the Earth as made up parts that connect together to form a whole, there is no sustainability in one part without the other. Instead, it is about the sustainability, or lack of sustainability, of the Earth as a whole. Based on the Gaia concept, there is a need to resolve issues in a more globalized manner while still acknowledging local dynamics. For example, when the situation in developing countries is not conducive to human habitability, the risk is not confined to just the Global South, but the Earth as a whole.

3.1   Gaia Theory and Contemporary Issues

The mid-twentieth century saw renewed studies of planetary habitability in effort to understand the origin and the existence of the Earth and life. Initially, the main motivations behind cosmology were to understand what exists beyond planet Earth and how the Earth looks like from afar. The exploration of the Earth and space was initiated by the Soviet space program in competition with the US National Aeronautics and Space Administration (NASA) during the Cold War. In 1961, Yuri Gagarin was the first human in space and between 1966 and 1972 NASA conducted the Apollo Program. To astronauts the Earth is a sphere floating in space. After living in orbit and viewing Earth from space, astronauts have expressed an increased connection to the Earth, its environment, and fellow human beings. Since 1961, the number of people whom have had the rare opportunity to space travel are few (i.e., just 565 as of 1 March 2020), but their feedback about what they saw has greatly inspired our understanding of the Earth itself. For instance, Samantha Cristofotetti has experienced one of the longest uninterrupted spaceflights (i.e., 199 days) expressed the more she stayed and viewed the Earth from space, the more her perception of humanity's time on the Earth evolved [26]. After visiting the International Space Station for the second time, Karen Nyberg said "[…] in the future, I would like to be more of an advocate for animal conservation. Every single part of the Earth reacts with every other part. It's one thing. Every little animal is important in that ecosystem. [Seeing the planet from above] makes you realize that, and makes you want to be a little more proactive in keeping it that way. If I could get every Earthling to do one circle of the Earth, I think things would run a little differently" [26]. When astronaut Leland Melvin viewed the Earth from space he felt that "we [humans] are genetically connected to this planet," [26]. Interestingly, the personal feelings of astronauts as observers of the Earth and the connection between human beings and nature should be a warning to better understand the Earth, life, and any custodial roles humanity has taken upon itself.

In an attempt to advance astronauts' views, various schools of thought and different theorists regarding the constellation of the Earth and life—including Charles Darwin's [27] theory of Darwinism which later advanced into Neo-Darwinism [28, 29]—explain biological evolution through transmutation of species and natural selection [28–32]. Yet, in the 1970s with the emerging concept of the Gaia hypothesis—the living organism of planet Earth floating in space—was proclaimed by Lovelock [25]. After Lovelock's engagement with NASA which heavily influenced his interest in the study of the universe and later co-developed by Lynn Margulis, a microbiologist, they named "Gaia" meaning "Earth" (i.e., in Greek the goddess of all life or mother Earth) as the founding concept and terminology of the Earth as mother nature.

According to Gaian theory, the Earth is a living organism and continues to evolve to achieve long-term stability. The current Earth's habitability is as a result of billions of years of evolvement, and self-regulating mechanization despite other forms of external perturbations such as increases in solar luminosity. The Earth, as stated by Lovelock [25], is a living creature floating among in space and made up of both organic (i.e., the biosphere) and inorganic components (e.g., the geologic environment, chemical composition, water, and surface pH). The idea of the Earth being an organism has been explained through stoic philosophy [25, 33, 34]. The interaction between organic and inorganic interact to form a living homeostatic entity often explained using concepts of natural theology. The Earth's biotic and abiotic components co-evolve as a single living and self-regulating system. The organic and inorganic components of the Earth are co-constituents; each component has its own system of self-regulation while each subsystem is a part of the other. The Earth as a living organism, with its own mechanisms that regulate species–species interaction and species-biosphere relations, creating harmony between organisms and the ecological sphere is its own bionetwork, in turn, aiding in the maintenance of their own relationship, self-survival, and Earth's self-regulation. As a result, this co-constitutiveness between Earth's elements is the composition of one ship. It is this co-constitutiveness that Christensen et al. [35] refers to as the tangled nature model (TNM) or tangled decision model [35–37]. During co-evolvement, the growth rate of every species depends on other species in the biosphere.

Gaian reasoning appears to fill the gaps left by Darwinism [27, 33] who believe that each species has been independently created and posits characteristics that fit the most favorable condition destined to live. To Lovelock [25], Earth consists of a complex interspecies relationship interconnecting the relationship between species and the environment [38]. Notwithstanding this, all forms of relationships within Earth's system are crucial, as such, a focus on human–human relations in understanding some of these dynamics are useful interplaying factors that warrant investigation. Gaian human–human interaction can create negative consequences that form a "system of [un]fitness" that can generate (un)fit mutantation. Note, the Gaia theory generates the idea of biological altruism through which the benefit to the entire Earth is founded on the altruistic actions of the individual. Gaian ideas appear somewhat complementary to those of Darwinian in that when the most admirable condition for survival occur a species is destined to live, hence its varying response to survive—including migration. The altruistic actions that Gaian theory is not necessarily meant for the benefit of the unfit, in their own ecology, maintain the whole organism—Earth—including the survival of philanthropists as a contingent part of human survival.

Gaian ideology explains that no species can grow independent of all others. This idea has been advanced through the logistic growth model in which no single species can form the majority of the population [39]. The growth rate is not intrinsically motivated but rather dependent on external factors (i.e., other species and the environment). However, in both cases, the growth rate is considered a linear function of interspecies interaction with no consideration for mutation among species. In real life, the world is very globalized and at the same time localized. The interconnectedness can be explained from the logistic growth model viewpoint, which sees species population as the constraint for its own growth and the TNM that widens the scope of species interaction to account for the effect of the total number of species in the whole system. Argued from anthropic principles in cosmology, the existence of Earth and life is explained by values of constants and variations (i.e., self-regulation)—relating to more or less a quantum mechanics outlook. Despite values of local constants to explain dynamics of the existence of life in different parts of the world, the coincidental effects of global constants of nature should not be underestimated. TNM would inform that issues in the Global South are not only an effect of interaction between people within the region but as a product of people's interaction within the global system.

3.2   Linking the Global North and the Global South for Sustainable Life on Earth

As informed by the Gaia theory, in order to understand the Earth's future stability, we case study human–human interaction and the foundation of sustainability of human settlement. To understand the Earth (i.e., one ship), the need to look at the developed and developing world (i.e., two classes) as distinct subsystems and co-constituents of human habitability of the entire world, metaphorically, is explored. The developed world is marked by internal stability. Developed countries have a well-functioning policy and legal framework. Governance is featured by the rule of law and advanced democratic settings in conjunction with high economic development processes (i.e., high human development index (HDI) scores). In addition, high industrial development and advanced technology (e.g., information and communications technology) have much improved human welfare. Hence, as a distinct subsystem, the developed world thrives on high levels of human welfare. Contrary, in the developing world, also known as the Third World, there are low economic development processes. The poor governance systems, with low levels of democracy, high poverty rates, wealth inequality, and low literacy levels, have resulted in the disillusionment of many people in this subsystem. These unfit socioeconomic systems yield more hopelessness manifesting in the form of prevalence of extreme poverty, radicalization, violent conflict, forced displacement, and irregular migration. Contrary to the Global North, the view of the Global South shows low human welfare, unsustainability, and, in some regions, misery (i.e., a state of catastrophe).

Balance in the ecosystem is maintained by optimum numbers. For instance, the numbers of gut bacteria are essential for the digestion of a ruminate. Within mutually beneficial systems, once they increase in numbers past the optimum threshold, they create risk not only to themselves but to the whole system. This occurs when species reduce habitability during their own interactions, making the system unstable and reducing total population. Based on Gaian ideology, negative environmental feedbacks are transmitted through human–human interaction to create negative externalities on the least competitively fit. However, many epidemic communities tend to see the Global South, Africa for instance, as a distinct subsystem where uncontrolled population growth rate is determined by people's interaction within the continent. Little effort is put to consider how disadvantages within a globalized world (i.e., the world as a holistic system) are avoided by individuals pendent on intellectual capability, social class, economic and political strength, and—eventually—passed over to the most unfit and poorest form of unavoidable condition (i.e., vulnerability to violent conflict and uncontrolled population growth). Species with low fitness will often reproduce [39]; however, the concept of "sequential selection" informs that periods with environmental improving species are longer compared to the periods of environmental degrading ones [40]. In reality, sequential selection is relevant in linking a degraded environment with low life expectancy.

Gaian ideology can also help enlighten the competition–cooperation–exclusion nexus among human beings [33]. One of the most outstanding global divides exist within social, economic, and political development between the Global North and the Global South. Informed by the Gaian notion that species increase their fitness for survival is relative to (socio-)environmental factors [41]. Population in the developed world and upper- and middle-class in low-income countries tend to converge this rationality (i.e., their socioeconomic environmental preconditions) of population planning and fertility preferences as a prerequisite for fitness [42]. In contrast, the poorest and marginalized can respond in two ways. First, there are those who increase in numbers due to lack of information or unmet social needs despite information.

Second, in most multi-ethnic and emerging democracies, rule is by the majority (i.e., not democratic governance defined by equity) in which tyranny of numbers where one or a combination of few social groups or ethnicities form the majority population and dominate power for their own benefit while excluding the other (i.e., often many) social groups that form the minority (i.e., in the same socio-political environment). In this case, to seek inclusion uncompetitive people can calculatively build political fitness by increasing numbers to the counter dominance. This same mechanism is accompanied by conflict and violent struggle for inclusivity and justice. These are more or less (in)voluntary strategies that Dutreuil [34] in expounding the Gaia theory refers to as "geoengineering techniques" relating to the human population and demographic change. In explaining these differential components between the developed and the developing world, from a Gaia hypothesis point of view, Earth as a living organism in which if one part is dysfunctional the organism as an entity risks survival. Despite high levels of development in the Global North, it is quite difficult to talk of sustainability without considering improved human welfare, prosperity, and sustainability of people in the Global South [43]. As a case scenario for the Global South, we explore Africa and examine the inheritance of conflict this continent endures.

3.3 Historicity, Mutation, and Inheritance of Conflict in Africa

In TNM, the relationship between the core-cloud and smaller-clouds can generate "unfit mutantation" within the "system of unfitness." Most of the least developed countries are featured by fragility or "system unfitness" and represent the total sum of fitness of all extant individuals. System fitness or unfitness can be inherited [36]. However, inheritance does not halt the co-evolution process of unfitness itself. Again, the two are co-constitutive processes within a system. Take for example, the current conflicts in Africa which form the major causes of forced displacement regionally and irregular migration out of the continent. Current violent conflicts are a result of a long co-evolution of historical fragility, inherited from one generation to another. The meta-issues that led to the emergence of violent conflicts in Africa in the 1960s remain to be the root causes of the current conflicts, only that the nature of conflicts themselves have largely mutated into differing forms and degrees. The linkage of Gaian ideology of mutation and inheritance of systems of fitness can be connected with the concept of post-modern conflicts, complexity theories of conflict resolution [44–47], and social theory that presents multidimensional understanding and revisits historicity of the root causes of fragility and conflicts throughout Africa. Argued from complex theories, mutation of systems of conflicts occurs because actors usually have different interests and continually change tactics in achieving those interests. This generates a changing, complex interplay between actors within a deconstructive setting ranging from ethnic cleansing, use of mercenaries, banditry, civil war, electoral violence to radicalization and terrorism [48]. To comprehensively address some of the global issues like irregular migration, the historicity of the current conflicts and fragility in migrants' home countries is often misplaced. There is the need to look back and reconstruct a proper historical-developmental agenda in an effort to look forward to developing strategies to achieve sustainable development goals (see Chap. 4 on sustainable land reform for further information). Forced displacement (i.e., either forced by violent conflict or extreme poverty) and associated irregular migration raise the alarm on the conditions of human life and habitability in many parts of the developing world. However, based on the Gaian concept of co-constitutiveness of the Earth and human life, seeing it as a problem of developing countries is a reductionist and fallacious understanding of the issues. Earth needs to be studied "as a whole" or "holistically" [25]. Therefore, the havoc that people run away from should be understood as a risk to human beings (i.e., to everyone) [43].

Electing barriers between the developed world and developing countries is increasingly becoming the trending strategy of managing irregular migration. It is an elusive idea that there is a possibility of establishing or maintaining a sustainable "subsystem of high(er) development" on a failing Earth. Like astronauts felt about the Earth when they observe it from space—the Earth, the environment, and human beings are all connected. To achieve sustainability, the Earth—life and human habitation included—should be seen as connected parts examined in multidimensional perspectives that acknowledge historicity of patterns of human settlement. For example, asylum management cannot be complete without input from environmental conservation agencies and the work of environmental promotion—and conservation will not be sustainable—when displaced persons or other vulnerable communities are excluded. To understand key facets of what has made the Global North system fit an examination of some of its key proponents can be compared. We highlight economic measures as the key subsystem difference which palpably forms, and supports, domino-like effects in conjunction with other measurements such as the environment, societal stability, and politics.

4   The Economy: A Subsystem Indicator

For millennia, economies have complemented the rise and fall of civilizations. Correspondingly, people undertake much of their decisions based on economic factors. This inert economic mechanism has evolved over millennia from primitive communities, bartering, slavery, feudalism, the market economy, communism, and other hybrid-related systems. Historicity of economics and the amount of wealth correlate with the majority of human history for the majority of human beings as impoverished (i.e., primarily due to scarcity and rulership). Historicity of the aristocracy (i.e., ruling class) has dictated much of the suppressive functionality of the masses. Suppression has primarily kept the masses down (e.g., via enslavement) and ignorant (e.g., being uneducated and uninformed). Institutional frameworks which principally supported and favored the aristocratic-end of society, eliminating any chance for welfare for the remaining population, has benefitted only the few for much of human history.

The foundation and possibilities for entrepreneurialism (i.e., shaped after agrarianism and mercantilism) and freedom in the decision-making process expanded human capacity to acquire wealth. The first profiteers to have grown out of the slave-class and feudal-system to create the earliest forms of capitalism occurred in Western Europe—a significant region of the contemporary Global North. Historic conditions set in the Global North—i.e., technologies of mass production, the ability to independently and privately own and trade goods and services, a working class willing to sell their labor, a legal commercial framework, a physical infrastructure to allow for the circulation of goods (i.e., on a large-scale), and security for private accumulation—created the base necessities for subsystem differentiation, something much of the Global South, still today, does not possess. The obstacles for the development of capitalist-based markets are therefore less technical and more social, cultural, and political.

One of the building principles for a successful economy was founded from mercantilism. The idea is connected with self-sufficiency of a country's production, avoiding the importation of goods and services from abroad (i.e., if necessary, enacting heavy duties on foreign goods) and promoting the transfer of gold and silver (i.e., wealth) to a country. This policy meant limited economic contact, due to a separation from economic competition, creating monopoly-oriented barriers for economic development. Conversely, in opposition to mercantilism, the idea of laissez-faire or economic freedom for business activities came to the forefront. The laissez-faire system was the autonomy businesses required to sprout and enlarge a new economic order. The public sector (i.e., government) reduced its activity [49] and space for individual economics expanded. At length, borders between countries would be open for free trade and capitalism unearthed.

In the eighteenth century, Smith [50] published "Inquiry into the Nature and Causes of the Wealth of Nations" which put under scrutiny the economic system, the principles of the market economy, the rules between the private and public sector, the individual, the needs to separate the public sector from the market, and the freedom to create and compete. These principles built the basis for today's market economy system (i.e., what became in opposition to mercantilism at the time). From the eighteenth century, signs of the First Industrial Revolution caused new technological change (e.g., machinery in textile industries, steam and internal combustion engines, and large-scale factories). This process developed into the nineteenth century in which some parts of the world (i.e., mostly today's Global North) started to practice the principles of the market economy (i.e., private property domination, no price and contract regulation, maximization of profits with minimized cost in business activity, freedom in doing business, as well as the free market to decide on location and development activity)—i.e., limiting big government and income-based redistribution.

As the system of the market economy outwardly opened, it gave way to increases in growth rate and growth income per capita. It also created the possibility to decrease poverty as the majority of people, at the time, worked in agriculture. Until the twentieth century, the market economy and Second Industrial Revolution influenced gross domestic product (GDP) per capita in Western Europe and the USA which increased twenty-five times during the period from the nineteenth century through to the beginning of the twenty-first century. Western Europe and the USA were regions with consistent market-oriented policy (i.e., with periodic changes from intervention to the market economy). During this period, income per capita increased from about US $100 to US $2500 per month [51]. The alternative to this system grew out communist and authoritarian regimes in which centralized planned economics (i.e., a public monopoly-based system) came to fruition. In central-planned economies, central decision-making processes dominate public property, rely on a high level of income redistribution, and economic goals connect directly with government objectives.

Considering the World Bank's [52] database on worldwide wealth and its classification of GDP per capita, over the last two decades, differences in wealth illustrate alarming disparity (Table 1). Utilizing GDP per capita data, it is evident that high-income countries (e.g., Australia, Ireland, Luxembourg, Monaco, Norway, Switzerland, and the USA) are market economy-friendly and all a part of the Global North. Middle-income countries (e.g., Angola, Argentina, Belarus, Brazil, Colombia, Egypt, India, Kyrgyzstan, and Nigeria) are from all continents, except Australia, in which not all market economic standards are fulfilled. Finally, the group of low-income countries usually located in Africa and Asia—exemplified by South Sudan, Somalia, Niger, Burundi, North Korea, and Tajikistan—the market economy exists only at a very limited scale.

Moreover, a supplementary indicator such as the HDI can provide a record that considers quality of life and not solely quantity (i.e., value of production). HDI is made up of life expectancy at birth, expected years of schooling, mean years of schooling, and gross national income per capita. The latest available HDI report from 2017 lists 189 countries. The top-end (i.e., leaders) of the report are Norway, Switzerland, Australia, Ireland, and Germany (i.e., all from the Global North), while the bottom-end are Burundi, Chad, South Sudan, Central African Republic, and Niger (i.e., all from the Global South) [53]. This evaluation, connected with the previous argument on the market economy, stresses the successes of the Global North and

Table 1 Worldwide GDP per capita (US $), 2000–2018

	High-income countries	Middle-income countries	Low-income countries	Average world income
2000	25,593	1272	314	5491
2004	32,098	1675	350	6811
2008	40,417	3382	597	9413
2012	41,867	4823	733	10,589
2016	40,763	4817	737	10,248
2018	44,705	5484	811	11,296

Source World Bank [52]

the failures of the Global South. Economic and social systems' need for effective institutional frameworks to stabilize expectation and business activity—in conjunction with formal and informal procedures to bring opportunity for legal businesses to thrive, be protected under the law, and provide private property rights—interplay the significance of human and social capital influences. Human capital-oriented factors (e.g., level of education and qualification) must be connected to social capital (i.e., culture, religion, world view, and history of nation). Factors that divide societies into contradictory cultures (i.e., individual–collective, monochronic–polychronic, and protransaction–propartnership) influence economic output and society at large [54]—the result of these factors interact regionally as well as internationally in terms of development diversification. One common governing intervention technique includes cohesion policy where wealthy countries and regions assist the least-favored in reducing disparities (see Chap. 8 for an example on European Union (EU) cohesion policy) [55–57].

Economic disparities is one measurement between the Global North and Global South. In terms of the Gaia theory, if one part of the Earth malfunctions, it endangers the entire Earth system including the safer and well-functioning areas. Accordingly, the sustainability of the Global North or the unsustainability of the Global South, as distinct subsystems, exists in unison. All parts (i.e., both biotic and abiotic) are entangled, and the world is globalized. When the Earth is observed from space, the notion of sustainable or unsustainable Earth is seen as a whole. The rising irregular migration from South America to the USA and from Africa as well as the Middle and Far East to Europe does not only indicate unsustainable human habitation in the areas where people are fleeing, but also the unsustainability of human habitability for the entity of the Earth. The guarantee of sustainable human life in the Global North is thus dependent on how it works with the Global South's development of sustainable human habitation and ability to attain a shared vision of prosperity. In practice, this calls for development agents expanding their scope to operate and capture entangled and globalized dynamics of human challenges in the Global South while addressing other insecurities such as the environment. The underlying aim of this book is, hence, to increase sustainability-based know-how and force a human–nature relations dialogue between the philosophical, economic, and urban arguments human beings toil with in modern society.

5 Synopsis of the Chapters

Chapter 1 "Human-Nature Relations: The Unwanted Filibuster" by Giuseppe T. Cirella, Samuel W. Mwangi, Andrzej Paczoski, and Solomon T. Abebe is a philosophical and introductory overview of human–nature and human–human relations. As the introductory chapter, specific Global North and Global South arguments interwoven with the notion of human settlement and differentiation between the two are explored. The objectives of the book as well as a synopsis of the individual chapters piece together and bridge an outline for the three parts: sustainability in conjunction with understanding and insight, economic evaluation, and urbanization and related strategies.

Chapter 2 "Sociocultural-Carrying Capacity: Impact of Population Growth in Rapa Nui" by Piroska Ángel and Kay Bergamini explores population increase for the Rapa Nui local community and establishes measures of management to ensure its sustainability. A governing bill has been developed to regulate the permanence, residence, and transferability to the Chilean territory to assist in calculating carrying capacity. Measures to overcome these concerns were administered by the Pontifical Catholic University of Chile in which impact from an increase in population (i.e., residents and tourists not belonging to the local ethnic group) is the fundamental target. As a tool-based approach, the bill oversees the management of livelihood, preservation and safeguarding of local identity, heritage safe-keeping, and social coexistence.

Chapter 3 "Territorial Integration of Foreigners: Social Sustainability of Host Societies" by Federico Benassi and Alessia Naccarato evaluates foreign population as a structural trait in Italian society. Territorial integration is a key factor in social sustainability. Mixed couples, an emerging phenomenon in Italy, lead to a change in social space and residential geography of the local environment. A theoretical reflection on the importance of territorial integration of foreigners and an assessment of the dimensional effects presented to the host societies' social cohesion, as well as an empirical application to examine the relationships between foreigners' residential integration and mixed couples, is studied. Territorial integration of foreigners assesses the linkages to the local, social environment in context of multi-segregation in conjunction with growth of mixed-race couples.

Chapter 4 "Sustainable Land Reforms and Irregular Migration Management" by Samuel W. Mwangi and Giuseppe T. Cirella expands upon land-related issues, directly and indirectly, affecting instability, conflict, and violence in Africa. Root causes of human displacement and irregular migration throughout the continent relationally connect historical land injustice, mitigating marginalization, and implementation of land reform. Specific land reform legislation within the EU-Africa partnership on migration management reviews major policy documents as well as concepts linking human security to land security and relational ties to socio-ecological vulnerability and resilience.

Chapter 5 "Role of the International Ecological Network, Emerald, in the Western Balkans' Protected Areas" by Tea Požar and Giuseppe T. Cirella assesses the role in controlling, preserving, and protecting nature at the European level. Research is focused on the Western Balkan countries and nature protected areas on the borders of these countries. Historical overview of the political situation before the Yugoslavian war is considered and consequences of this event which left nature protected areas are reviewed. Cross-border cooperation among many countries, cities and villages, or even whole regions is very dependent on cooperation between geographical entities. By enlarging Western Balkan countries into the EU, it is important to observe the Areas of Special Conservation Interest changes to the Emerald Network which presents a basis for addressing future EU-level Natura 2000 sites.

Chapter 6 "How Efficient is Urban Land Speculation?" by Bedane S. Gemeda, Birhanu G. Abebe, and Giuseppe T. Cirella explores mathematical formulation for best urban land speculation practice and present planning challenges of contemporary urbanization and settlement patterns. Novel research is specific to the sub-Sahara where few studies have explicitly examined the policy forces driving this part of the world. This chapter analyzes the urbanization and policy driving land speculation, economic role of land speculators, opportunity of land speculation, and motives behind land speculation in the city of Shashemene, Ethiopia. The research encompasses an attempt to discover the process of land speculation in peri-urban and urban land by examining principal actors. Findings illustrate the societal costs in a typical city are increased by about 5% to 11% as a result of speculative increases in the value of urban land.

Chapter 7 "Land Use Change Model Comparison: Mae Sot Special Economic Zone" by Sutatip Chavanavesskul and Giuseppe T. Cirella looks specifically at the development of several basic infrastructure-related projects and public sector mega department stores under construction in Thailand. To date, land use change plays an important part in influencing the area. This case study uses land use change and prediction modeling over a 20-year period into the future, i.e., 2028 and 2038, by comparing the cellular automata-Markov model and Land Change Modeler methods. Predictive results show similar findings from both methods. It is determined that these methods can assist in properly monitoring impact on society, environment, security, and public health.

Chapter 8 "Cohesion Policy for Europe 2020" by Andrzej Paczoski, Solomon T. Abebe, and Giuseppe T. Cirella observes elements of the EU's integration processes. Mechanisms show the efficiency of European policy convergence and its elimination of development inequalities between Member States (MSs). The Europe 2020 Strategy introduced new challenges for MSs through strategic goals piloted by target figures. Cohesion policy, as a tool, is used to harmonize development and decrease development disparities. Europe 2020 targets are examined from 2011 to 2018 and highlight, at the national-level, lack of effective institutional framework, transport infrastructure, education, and innovation policy.

Chapter 9 "Evaluating Green Infrastructure via Unmanned Aerial Systems and Optical Imagery Indices" by Matjaž N. Perc and Giuseppe T. Cirella pieces together research on remote sensing techniques for evaluating green infrastructure via unmanned aerial systems (UAS). In terms of urban sustainability, this approach can be applicable when, otherwise expensive or up-to-date imagery is unavailable. This chapter illustrates the use of UAS as an adjunct to best urban planning practice and landscape infrastructural design by collating environmental applications for green urban infrastructure mitigation. The study area, located in Slovenia, assembles bottom-up imagery to produce a UAS-based flood map from predicted and potential rainfall. The utility of UAS, as a complementary environmental monitoring tool, can be used to direct natural and urban systems synchronicity.

Chapter 10 "Urban Sustainability: Integrating Ecology in City Design and Planning" by Alessio Russo and Giuseppe T. Cirella evaluates urban sustainability in conjunction with ecosystem services and biodiversity. Sound urbanized environments will need to relate urban planning and urban design processes. Adaptive urban know-how is at the root of this chapter in which a number of exploratory concepts and notions are put forth with the intention of creating dialogue between ecosystem services and human well-being (i.e., through concerted ecological, social, and economic action). The chapter reviews a number of cases and concludes with background research in properly developing sustainable models and tools.

Chapter 11 "Urbanization and Population Change: Banjar Municipality" by Agus Supriyadi, Tao Wang, Shanshan Chu, Tianwu Ma, Raden G. Shaumirahman, and Giuseppe T. Cirella seeks to show how the increase in concentration of people moving to cities impacts land use. A case study examines Banjar Municipality, a new autonomous city, in Indonesia by comparing land use change between 2006 and 2016. Results show an increase in urban area usage and a decrease in agricultural land. The underlying research focuses on constraints to facilitate land use within an urban corridor setting.

To help with scientific jargon and terminology, a glossary of terms and index have been included at the end of the book.

References

1. Schultz PW (2002) Inclusion with nature: the psychology of human-nature relations. Psychology of sustainable development. Springer, US, Boston, pp 61–78
2. UN (2019) World urbanization prospects: the 2018 revision. United Nations, Department of Economic and Social Affairs, New York
3. Redman CL (1999) Human impact on ancient environments. University of Arizona Press, Tucson
4. Ellis EC (2011) Anthropogenic transformation of the terrestrial biosphere. Philos Trans R Soc A Math Phys Eng Sci 369:1010–1035. https://doi.org/10.1098/rsta.2010.0331
5. Ruddiman WF (2013) The anthropocene. Annu Rev Earth Planet Sci 41:45–68. https://doi.org/10.1146/annurev-earth-050212-123944
6. Barnosky AD (2014) Palaeontological evidence for defining the anthropocene. Geol Soc Lond Spec Publ 395:149–165. https://doi.org/10.1144/SP395.6
7. IPCC (2013) AR5 climate change 2013: the physical science basis—IPCC. Working Group I of the Intergovernmental Panel on Climate Change, Geneva
8. Soulé ME (1985) What is conservation biology? Bioscience 35:727–734. https://doi.org/10.2307/1310054
9. Doak DF, Bakker VJ, Goldstein BE, Hale B (2014) What is the future of conservation? Trends Ecol Evol 29:77–81. https://doi.org/10.1016/J.TREE.2013.10.013
10. Vucetich JA, Bruskotter JT, Nelson MP (2015) Evaluating whether nature's intrinsic value is an axiom of or anathema to conservation. Conserv Biol 29:321–332. https://doi.org/10.1111/cobi.12464
11. Cafaro P, Primack R (2014) Species extinction is a great moral wrong. Biol Conserv 170:1–2. https://doi.org/10.1016/j.biocon.2013.12.022
12. Kareiva P, Marvier M (2012) What is conservation science? Bioscience 62:962–969. https://doi.org/10.1525/bio.2012.62.11.5
13. Ponting C (2007) A new green history of the world: the environment and the collapse of great civilizations. Penguin Books, London

14. Ceballos G, Ehrlich PR, Barnosky AD et al (2015) Accelerated modern human–induced species losses: entering the sixth mass extinction. Sci Adv 1:e1400253. https://doi.org/10.1126/sciadv. 1400253

15. Perman R (2011) Natural resource and environmental economics, 4th edn. Pearson Addison Wesley, Gosport

16. Kleijn D, Winfree R, Bartomeus I et al (2015) Delivery of crop pollination services is an insufficient argument for wild pollinator conservation. Nat Commun 6:7414. https://doi.org/10.1038/ncomms8414

17. Plumer B (2016) What bees can teach us about the real value of protecting nature—Vox. In: Vox. https://www.vox.com/2015/7/6/8900605/bees-pollination-ecosystem-services. Accessed 22 Jul 2019

18. Chan KMA, Balvanera P, Benessaiah K et al (2016) Opinion: why protect nature? Rethinking values and the environment. Proc Natl Acad Sci 113:1462–1465. https://doi.org/10.1073/PNAS.1525002113

19. Díaz S, Demissew S, Carabias J et al (2015) The IPBES conceptual framework—connecting nature and people. Curr Opin Environ Sustain 14:1–16. https://doi.org/10.1016/J.COSUST. 2014.11.002

20. Piccolo JJ (2017) Intrinsic values in nature: objective good or simply half of an unhelpful dichotomy? J Nat Conserv 37:8–11. https://doi.org/10.1016/J.JNC.2017.02.007

21. Spash CL (2015) Bulldozing biodiversity: the economics of offsets and trading-in nature. Biol Conserv 192:541–551. https://doi.org/10.1016/J.BIOCON.2015.07.037

22. Steffen W, Richardson K, Rockstrom J et al (2015) Planetary boundaries: guiding human development on a changing planet. Science (80-) 347:1259855–1259855. https://doi.org/10.1126/science.1259855

23. Callicott JB (2014) Thinking like a planet: the land ethic and the earth ethic. Oxford University Press, Oxford

24. Shoreman-Ouimet E, Kopnina H (2016) Culture and conservation: beyond anthropocentrism. Routledge, New York

25. Lovelock J (1979) Gaia: a new look at life on Earth. Oxford University Press, Oxford

26. National Geographic (2008) These astronauts saw earth from space: here's how it changed them. In: National Geographic Magazine. https://www.nationalgeographic.com/magazine/2018/03/astronauts-space-earth-perspective/. Accessed 22 Jul 2019

27. Darwin CR (1859) On the origin of species by means of natural selection, or, the preservation of favoured races in the struggle of life, 1st edn. John Murray, London

28. Costa JT (2014) Wallace, Darwin, and the origin of species. President and Fellows of Harvard University, Cambridge

29. Flew A (2018) Darwinian evolution, 2nd edn. Routledge, London

30. Sclater A (2006) The extent of Charles Darwin's knowledge of Mendel. J Biosci 31:191–193. https://doi.org/10.1007/BF02703910

31. Lopreato J, Crippen TA (1999) Crisis in sociology: the need for Darwin. Routledge, New York

32. Padian K (2018) Origins of Darwin's evolution: solving the species puzzle through time and place. Syst Biol 67:741–742. https://doi.org/10.1093/sysbio/syy016

33. Cazzolla Gatti R (2018) Is Gaia alive? The future of a symbiotic planet. Futures 104:91–99. https://doi.org/10.1016/J.FUTURES.2018.07.010

34. Dutreuil S (2018) James Lovelock's Gaia hypothesis: "A New Look at Life on Earth"... for the life and the earth sciences. In: Dreamers, visionaries, and revolutionaries in the life sciences. University of Chicago Press, Chicago, pp 272–287

35. Christensen K, Di Collobiano SA, Hall M, Jensen HJ (2002) Tangled nature: a model of evolutionary ecology. J Theor Biol 216:73–84. https://doi.org/10.1006/JTBI.2002.2530

36. Laird S, Jensen HJ (2006) The tangled nature model with inheritance and constraint: evolutionary ecology restricted by a conserved resource. Ecol Complex 3:253–262. https://doi.org/10.1016/j.ecocom.2006.06.001

37. Arthur R, Sibani P (2017) Decision making on fitness landscapes. Phys A Stat Mech its Appl 471:696–704. https://doi.org/10.1016/J.PHYSA.2016.12.068

38. Lovelock JE, Margulis L (1974) Atmospheric homeostasis by and for the biosphere: the Gaia hypothesis. Tellus 26:2–10. https://doi.org/10.3402/tellusa.v26i1-2.9731
39. Arthur R, Nicholson A (2017) An entropic model of Gaia. J Theor Biol 430:177–184. https://doi.org/10.1016/J.JTBI.2017.07.005
40. Ford Doolittle W (2014) Natural selection through survival alone, and the possibility of Gaia. Biol Philos 29:415–423. https://doi.org/10.1007/s10539-013-9384-0
41. Mustonen V, Lässig M (2009) From fitness landscapes to seascapes: non-equilibrium dynamics of selection and adaptation. Trends Genet 25:111–119. https://doi.org/10.1016/j.tig.2009.01.002
42. Odusola A (2018) Poverty and fertility dynamics in Nigeria: a micro evidence. SSRN Electron J 1–28. https://doi.org/10.2139/ssrn.3101818
43. Lenton TM, Latour B (2018) Gaia 2.0. Science (80–) 361:1066–1068. https://doi.org/10.1126/science.aau0427
44. Hughes S (2004) Complexity theory: understanding conflict in a postmodern world. Marquette Law Rev 87
45. Jones W, Hughes SH (2003) Complexity, conflict resolution, and how the mind works. Confl Resolut Q 20:485–494. https://doi.org/10.1002/crq.42
46. Baker M (2006) Translation and conflict: a narrative account. Routledge, New York
47. Demmers J (2016) Theories of violent conflict: an introduction. Routledge, London
48. Jackson R (2002) Violent internal conflict and the african state: towards a framework of analysis. J Contemp Afr Stud 20:29–52. https://doi.org/10.1080/02589000120104044
49. Keaney M (2017) Questionable interventions: the enduring myth of Laissez-Faire. Polit Stud Rev 15:404–414. https://doi.org/10.1177/1478929916646391
50. Smith A (1776) An inquiry into the nature and causes of the wealth of nations. Methuen, London
51. Tarchalski K (2016) Ewolucja instytucji kapitalizmu Zachodu. kapitalizm. Fakty i iluzje. Wydawnictwo Nieoczywiste, Rzeszów, pp 67–98
52. World Bank (2019) World Bank open data. In: World Bank. https://data.worldbank.org/. Accessed 27 Jul 2019
53. UNDP (2017) Human development index: United Nations Development Programme. http://hdr.undp.org/en/statistics/hdi/. Accessed 17 Jul 2019
54. Paczoski A (2014) Różnice kulturowe wobec wzrostu gospodarczego: Ze szczególnym uwzględnieniem przykładów państw UE (Translated from Polish: "Cultural differences and economic growth: with emphasized examples from the EU"). Polityka Gospod 22:91–118
55. Mohl P (2016) Empirical evidence on the macroeconomic effects of EU cohesion policy. Springer Fachmedien Wiesbaden, Wiesbaden
56. Molle W (2007) European cohesion policy. Routledge, London
57. Barca F (2008) An agenda for a reformed cohesion policy a place-based approach to meeting European Union challenges and expectations. Economics and Econometrics Research Institute, Brussels

Sociocultural-Carrying Capacity: Impact of Population Growth in Rapa Nui

Piroska Ángel and Kay Bergamini

Abstract A sustained population increase in Rapa Nui has been of a significant concern for the local community since the impact raised the need to establish measures of management to ensure its sustainability. To this end, a governing bill has been developed to regulate the permanence, residence, and transferability to the territory, as well as to promote technical instrumentation to calculate the carrying capacity of the island. A series of measures to overcome these concerns was administered by the Pontifical Catholic University of Chile of which this chapter investigates. A proposed methodology to measure the sociocultural-carrying capacity of Rapa Nui and monitor the impact generated from such phenomena, brought about by the increase in population (i.e., residents and tourists not belonging to the local ethnic group), is the fundamental target. Development of a tool-based approach, capable of managing livelihood, preserving and safeguarding local identity, heritage safe-keeping, and social coexistence in the insular territory is applied.

Keywords Carrying capacity · Sociocultural · Methodology · Rapa Nui

1 Introduction

As a result of new inhabitants attracted by economic and tourism opportunities, a demographic explosion in Rapa Nui (i.e., Easter Island), a Chilean island territory located in the Pacific Ocean 3800 km from the mainland (Fig. 1), constitutes a significant concern for the community and relating impacts on territorial integrity. The island community is taking measures to ensure sustainability and preservation of its cultural and environmental heritage. This action motivated the approval of a law that seeks to regulate the settlement, visitation, and transfer to Rapa Nui.

P. Ángel (✉)
Observatory of Cities UC, Pontifical Catholic University of Chile, Santiago, Chile
e-mail: paangel@uc.cl

K. Bergamini
Institute of Urban and Territorial Studies, Pontifical Catholic University of Chile, Santiago, Chile
e-mail: kbergani@uc.cl

© Springer Nature Singapore Pte Ltd. 2020
G. T. Cirella (ed.), *Sustainable Human–Nature Relations*,
Advances in 21st Century Human Settlements,
https://doi.org/10.1007/978-981-15-3049-4_2

23

Fig. 1 Location of Rapa Nui, Chile

The highest inter-census growth occurred between 2002 and 2017, with an increase of 104% representing the highest growth rate at the national level [1, 2]. Said law points out the need for a technical instrument to calculate the carrying capacity of the territory, taking a series of measures based on defined thresholds. In this context, the Undersecretary of Regional and Administrative Development entrusted the Pontifical Catholic University of Chile the task of developing this instrument used in framing the governing bill.

Initially, in instrumenting the law, demographic, environmental, infrastructure, and sociocultural variables were observed, followed by bibliographical and methodological approaches as well as piecing together defined and identified sociocultural-carrying capacity levels. The research proposes a methodology to measure and monitor the sociocultural-carrying capacity in Rapa Nui, largely based on the phenomena of increased population and effect on social relationships between residents and foreigners. The ultimate goal has been to provide an instrument with the capacity of guiding a management model in order to preserve and protect the local identity, heritage, and social coexistence of the territory. The thesis of the research project is identified under the Easter Island Demographic Carrying Capacity Study, developed by the Pontifical Catholic University of Chile.

The context in which this law emerged dates to 2007 in which constitutional reform qualified Rapa Nui, along with the Island of Juan Fernández, Chile, as a Special Territory, due to their fragility and ecosystem vulnerability, as well as unique cultural and natural heritage [3]. The objective of promoting the protection, development, and precaution of the island territory is proposed. A series of legislative and administrative actions, among the most significant, regulate the increase of residents based on restrictions and limitations due to a series of noted effects [3]. In accordance with President Michelle Bachelet signed into law on 30 April 2016 the "Residence, Permanence, and Transfer to the Special Territory of Easter Island," which included indigenous consultation (i.e., with a 97.7% approval rating) [4], on March 23, 2018, Law 21-070 that "Regulates the exercise of the rights to live, stay, and move to and from the Special Territory of Easter Island" was further decreed via the government's official gazette.

2 Problem Statement

Tourism, as an economic activity, has been questioned by the adverse effects and friction it can cause on a destination, despite the importance it can exercise on the local economy and improvement of the quality of life of local residents [5, 6]. The growth and inadequate operation of this action can end the patrimonial and cultural richness of a territory [7], which indicates costs that will be reciprocally absorbed, be it society, local community, private sector, or individuals [8]. To promote the principle of sustainability, the management of negative impacts, such as the transformation of place, effects of temporary overcrowding or demographic increase, loss of biodiversity, exclusion of the community, multinational monopolies, and among other effects, can be irreversible particularly when fragile ecosystems intervene in vulnerable societies [5, 8, 9]. This problem is evident on Rapa Nui, namely it is a product of archaeological heritage, preservation of living culture (i.e., intangible heritage including not only traditions inherited from the past but also contemporary rural and urban characteristics of various cultural groups) [10], its biodiversity, and geography as one of the most important tourist destinations in Chile [11, 12]. The singularity generated by the combination of all these variables has promoted a tourist explosion in the last three decades, affecting the permanent economic growth and improvement of inhabitants' income, which undoubtedly indicates significant economic benefits to the community [13]. Nonetheless, disadvantages and vulnerabilities of this insular condition exist, including: ecological fragility, limited resources, and isolation to markets. Shockingly, some studies indicate Rapa Nui may be showing signs of a tourism collapse due to the deterioration and mismanagement generated to its archaeological, natural, and cultural systems [10, 13]. Experience from around the world demonstrates sustainable management of tourist destinations, through tools and guidelines, focus heavily on environmental and economic components, often promoting quality of the tourist experience in order to ensure the tourism market [14, 15]. For this reason, studies that reveal the impact exerted on communities, their culture, and way of life, lack prominent tourism planning where social relationships, perception, experience, and local expectation are involved [14–16]. The importance of their inclusion is fundamental, even though harmful social and cultural effects—unlike economic ones—occur gradually and discretely, since they are permanent and with little or no opportunity of reversing them [17]. Among its effects, tourism represents an essential contribution of economic means, making it possible for migratory flows and augmentation of resident populations to generate aftereffects territory-wide [18, 19]. Sociocultural aspects are loss of traditional values and cultural diversity, social conflicts, cultural clashes, and social and cultural maladjustment [8, 20, 21].

Once the culture has become valuable, sense of identity is reinforced and the community can reevaluate its self-esteem and allow for the maintenance of some local customs and uses. A multicultural exchange between people from different countries and cultures is encouraged in an enriching manner, as well as heritage conservation promoted due to tourism income and international relevance generated by knowledge of its value [21]. This duality presents the tourist activity as opportunistic versus total collapse, from visitor-to-local interaction intervening on environmental, social, and economic scales to generate community breaks and alliances that are acceptable or not; however, mutual dialogue suggests a high degree of integration between the host and guest as desirable. For the case of Rapa Nui, one of the consequences (i.e., adverse effects) of tourism consolidation has been the population explosion, a product of the arrival of new inhabitants attracted by the economic possibilities offered by the activity [22]. For this purpose, an essential part of this chapter elucidates and responds to this problem and its manifestation in different territorial components derived from the concept of carrying capacity [23].

One of the most common definitions of carrying capacity is the maximum population that can support a given habitat without permanently damaging the productivity of the ecosystem on which the population depends [24]. From this definition, the concept has a close relationship with ecology, yet the idea since its initial postulation has been evolving as a result of transformations in political and epistemological contexts [25]. In this way, the concept has developed a variety of perspectives to include ecological areas, urbanism, political frontiers, and psychosocial economics and cultural changes [25, 26]. When the capacity of local systems overflow—as with the case in Rapa Nui's population explosion—a decline, not only in vital ecosystem services occurs but also in local heritage (i.e., material and immaterial) and quality of life. From this perspective, the objective of planning and management is to establish a life cycle, as sustainable as possible, medium- and long-term at a local and global scale, and for this, it is necessary to work with load limits in different territorial components in an integral way. The sociocultural components are fundamental to establish these limits and divert adverse effects on local identity, cultural evolution, and quality of life at the community level [27]. The carrying capacity concept can take charge of sociocultural sustainability insofar as it incorporates respect for cultural authenticity of host communities, the conservation of their cultural and architectural assets, and their traditional values as well as promoting intercultural tolerance. Thus, the sociocultural-carrying capacity responds to the effects that can be generated from the "touristification" of a territory on the sociocultural component, which can promote changes in traditional ways of life, changes in family structure and community organization, intercultural conflicts, and over-commercialization of arts and crafts that can lead to the loss of authenticity [25, 28]. Fostering positive sociocultural effects include: flourishment of a broad spectrum of artistic activities and cultural services, promotion of enriched exchange between people from different cultures, reactivation and development of local traditions, rescue of vernacular languages and dialects as well as oral culture, and the added-value of traditional collective knowledge [8].

On the other hand, it is necessary to recognize the dependence of Rapa Nui on tourism development as relevant, placing it in a fragile scenario to impacts of an overloaded population by way of deterioration of biophysical resources, overcrowding, and sociocultural impacts that immediately put in play the survival of the developed economic model. These effects have an inherent impact on the quality of life of all inhabitants on environmental, social, and economic components.

2.1 Case of Rapa Nui

According to demographic data, the population living on Rapa Nui has increased progressively over time. These results have been accompanied by a tendency both for the growth of the native population (i.e., the Rapa Nui ethnic group) as well as for a population that does not belong to it. In this case, the intercensal variation of the Rapa Nui population increased by 54% over the previous period, and in turn, the population not defined as Rapa Nui grew by 174%, which is particularly striking. Despite the increase in residents and immigration, it is necessary to add additional problems that arise from the floating, or tourist, population. Data indicates that annual tourist arrivals in the Rapa Nui National Park increased from 36,412 in 2007 to 60,856 in 2017 [29, 30]. This increase in population, both permanent and seasonal, has raised the demand from the Rapa Nui people to regulate the rights to reside, stay, and move to the territory. To understand the social dynamics behind this demand, it is possible to identify that continental migration to the island as the cause of the collapse of basic services (i.e., concurrent power cuts), increase in vehicle fleet, crime, waste, natural resource pollution, as well as loss of vernacular language due to the predominance of Spanish.

As such, the new governing bill seeks to establish specific measures that will enact certain limits of carrying capacity, defined by scientific instruments, aimed at improving and not exceeding land management capabilities. The formulation for calculating population-carrying capacity considers environmental characteristics, geographical conditions, current and potential land use, waste disposal, and sociocultural circumstance. To construct this instrument, a series of variables are identified to monitor certain problems. The implemented tool, not only yields the result of how many people fit in the territory but also inputs for its management, including future risk to tourism, the ecosystem, and community.

A lack of methodological contribution in the sociocultural dimension suggests an imbalance in empirical, sustainable management of the territory. What this means for Rapa Nui is any lack of importance given to the sociocultural issue from a scientific viewpoint should be considered when deciding on public policy. Policy that is territorial management and planning focused will need local community input to assist in creating a separation between the theoretical and empirical. As such, critical arguments must be considered, including: how to measure and monitor Rapa Nui's sociocultural-carrying capacity, how to monitor the impact on its sociocultural heritage (i.e., mainly from the increase in population both from residents and tourists), and what obligations do the local, ethnic group have over new-comers? Using these argumentative standpoints, the objective is to propose a sociocultural-carrying capacity measurement that monitors sociocultural impact on heritage due to the increase in foreign (i.e., non-local and ethic) population.

3   Theoretical Framework

As a result of economic growth accompanied by constant environmental and social deterioration, produced by ecological irrationality of dominant patterns of production and consumption, deep political and theoretical debates developed the advancement of sustainable development [19]. This promoted alternative models of development as a response to a certain awareness of impacts generated by economic systems that were proven to be unsustainable and incompatible with world reality [31]. From this, emerged, as in any other economic activity, arguments against the growth model of the tourism industry. The tourism industry has traditionally ignored environmental and sociocultural effects which have forced a rethinking of this relationship (i.e., between tourism, society, and development).

From the late 1980s to early 1990s, the idea of sustainable development of tourism disseminated (i.e., proper use of natural, cultural, and social resources) to guarantee their application for future generations [32]. Accordingly, sustainable development in tourism was drawn up to meet the needs of the present and future, fostering an economy that provides the necessary goods to the community while minimizing the adverse effects of the activity. The emergence of this awareness has given way to an industry that recognizes on the one hand, the positive effects from the creation of employment, increase of economic income, investment in the conservation of natural space, avoidance of emigration of local people, improvement of the economy, revaluation of culture, and commercialization of local products, and the negative impacts, in many cases, that outweigh the benefits [8, 32, 33]. In fact, tourism is recognized as one of the main instruments for poverty reduction, especially in developing countries and small island territories, but despite the obvious advantages, it has the potential to become self-destructive if the limits of growth are not respected with complementary policy decision making [34]. Several authors identify that the development of tourism contributes socially and economically to a territory; however, negative effects are as significant as the positive ones, due to the increase in the consumption of soil and water, increase in the production of waste, alteration of ecosystems, increase in population flows, loss of traditional values and cultural diversity, and inflation [8, 32–34]. It is extracted in the conclusion, that although the tourism industry has theoretically been recognized as a phenomenon of territorial scope, with impacts on the environmental, sociocultural, and economic variables manifested in the territory [35], it has been empirically developed mainly in the physical and economic dimensions.

Studies that expose and deepen the sociocultural dimension through the preservation of diversity, in its broadest sense, that is, preserving the system of values, practices, and social fabric of a community, as well as social relationships, perceptions, and conflicts, lack empirical landing.

3.1 Carrying Capacity

The discussion on carrying capacity has its beginnings in 1798 from the English economist Thomas R. Malthus who pointed out that population growth, if left uncontrolled, would eventually outpace the production of food. This basic theory of limitation imposed upon the environment would restrict the number of living organisms, including human beings, by way of limits of population growth achieved via balance. This theory was improved by Pierre F. Verhulst in 1838, which he labeled human-carrying capacity—commonly applied within demography. Later, another perspective installed the notion of ecological-carrying capacity which specified a biophysical limit to the environment as well as the concept for the management of fauna by establishing the maximum number of animals that could graze in an area without damaging soil or grassland production [36, 37]. Subsequently, from the 1960s and early 1970s, the concept is used in the field of applied ecology and human ecology, with a focus on maximum population size via socioeconomic scales based on maximum level of resource use, or capacity, for environmental assimilation. This provided information on the relationship between human activities and the environment [38]. Among these emerging approaches, much was dedicated to calculating the number of people a natural area could sustain without destroying essential, ecological characteristics. As such, interest primarily within the circles of American academia began to associate this concept with tourism planning [26, 28, 39–41]. However, the concept soon became a part of the overarching notion of the three pillars of sustainability: ecological (i.e., biophysical and the environment), social (i.e., cultural and human), and economic. Within this sustainability paradigm, the concept of carrying capacity became intertwined and remains to this day.

To date, the tourism-oriented side of the concept's guidelines has focused on works such as Miguel Cifuentes, among them the "Determination of the tourist change capacity of the visit sites of the Galápagos National Park" from 1996, which mainly focused on the field of ecology in natural protected areas. Methodologies have also emerged for calculating carrying capacity in areas of archaeological protection and monumental assemblies with methods adapted to those of Cifuentes, however, essential shortcomings identify impacts to the community, both socioeconomically and socioculturally, despite the fact that the concept has increasingly been associated with a sustainability framework. Conversely, the literature confirms a series of drawbacks associated with the measurement of carrying capacity, including: variety of measurement patterns, the notion that it remains a dynamic concept, missing quantifiable measures, difficulties in predicting impact, solutions proposed by different experts do not always reach a common consensus, lack of acceptable environmental indicators, and subjectivity of certain parameters. The above makes it understandable that the tourist-carrying capacity is often treated in the literature as academic and theoretical with little empirical application or studies [26].

3.2 Sociocultural-Carrying Capacity

The concept of sociocultural-carrying capacity responds to the effects that can be generated from the "touristification" of a territory, which can produce changes in traditional ways of life, family structures and community organizations, intercultural conflicts, over-commercialization of arts and crafts, and, hence, loss of authenticity. Knowing and understanding these consequences are necessary for the planning of sustainable development so local communities and cultures are permanent beneficiaries and not victims of tourism [42]. Likewise, it is relevant to state that the evolution of the concept, especially when applied in complex tourist destinations, emphasizes or exposes results of shifting nature (i.e., from numerical limits toward the development of theoretical and conceptual frameworks) for territorial planning and management. One of these impacts is in the development of "tourism culture," a product of mixture or contact between host and guest cultures, which can imply both positive and negative impacts. Nonetheless, this mutual interaction can lead to the development of new and different cultural landscapes, hence, a tourist culture. As a result, this complexity intertwined with a lack of specific measurement systems of sociocultural-carrying capacity facilitates the need to construct a measurement system supported by best practices.

At the same time, it should be noted that negative effects on the sociocultural dimension tend to act on small and traditional communities [25] and manifest through, for example, the abandonment of traditional activities (e.g., fishing and agriculture), displacement of other economic activities due to the attractiveness generated by tourism, breakdown of the values of the host community, changes in the consumption patterns caused by the availability of imported goods, gradual loss of local identity, increase in criminal acts and inter-ethnic conflict, and over-commercialization of the arts that alter lifestyle, value, and customs of a community [28]. In turn, sociocultural indicators can identify positive effects of the manifestation of the tourist activity, including: feeling of pride, intercultural exchange between visitors and residents (i.e., as an exercise of understanding and appreciation of the other), and heritage conservation (i.e., music, dance, arts, clothing, and ceremonies). Based on the construction of indicators, information can be extracted with the aim of managing future risks to the community, its heritage, and whether the tourist destination will suffer. The result of these indicators is not the construct of their components, but the apparatus to improve decision making for a sustainable territory. For this, it is necessary to identify a set of indicators with selection criteria, including: topic selection, feasibility to obtain and analyze data, credibility and reliability of data, and clarity to understand users in conjunction with overall comparability over time

[43]. This can be applied as long as there is awareness for the need and relevance of applicable management instruments to local sustainability.

4 Methodology

Given the lack of literature on sociocultural-carrying capacity measurement systems, a search of variables and indicators was used to measure, evaluate, and monitor socio-cultural sustainability with reference to research studies developed for indigenous cultures and isolated territories. At the same time, the need to establish a quantitative indicator system was defined, with the objective of providing comparable information for the analysis and monitoring of variables that could be measured over time. The research was structured into three key stages (Fig. 2).

The first stage of the methodology involved the definition of a matrix of quantitative indicators to measure Rapa Nui's sociocultural-carrying capacity, broadly consisting of an extensive literature review in order to identify sociocultural criteria and variables with the purpose of developing a first conceptual approach. Then, socio-cultural effects of tourism development together with the demographic explosion in the territory were defined with respect to different criteria and variables identified,

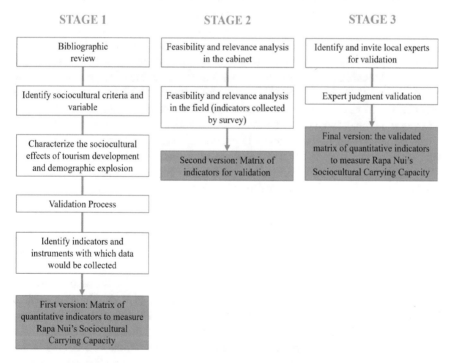

Fig. 2 Methodology to develop Rapa Nui's sociocultural-carrying capacity

through the review of secondary sources associated with the territory under study. In this way, an approach to the local problem was obtained.

Both methods (i.e., conceptual and experiential) are key inputs for a first-round selection of indicators. To prioritize the parameters, the Delphi method was utilized using a group of experts specialized in "carrying capacity" and "local reality" research. They were provided with both the conceptual matrix and sociocultural diagnosis, in order to support the matrix's validation process and system's functionality. With a validated conceptual matrix, we proceeded to identify indicators and instruments with which data would be collected, either through surveys, interviews, cadasters, or secondary sources, in order to establish the logistics of data collection.

The second stage consisted of screening the indicators according to the data and its relevance for which the following five filters put forth by the Food and Agriculture Organization of the United Nations [44] were used: (1) feasibility of collecting comparable and measurable information over time; (2) estimated quantity and quality of the data that can be obtained; (3) costs of information gathering; (4) relevance of the indicator to measure the objective; and (5) reliability of available information. A second version of the measurement system was obtained from this exercise.

Although the matrix of indicators had already gone through more than one selection filter, a third stage of validation with local experts was performed. The methodology proposed for this stage was based on the "expert judgment validation" approach. The experts were selected based on their knowledge and performance in different activities of Rapa Nui's sociocultural dimensions, as well as their availability and willingness to participate. This exercise eliminated those variables and indicators that were considered irrelevant for the case study and added those that were identified as essential. In turn, it was important that these changes were consistent with validity and reliability criteria in order to promote the quality of the instrument. The information was collected during a group workshop in which a level of consensus was applied to make the decision on how the final matrix of indicators would be structured. A breakdown of the methods of data collected are as follows: (1) survey: questions associated to the indicators collected through the "carrying capacity survey" instrument, (2) interview with key figures (i.e., gathering of expert information), (3) secondary source (i.e., information available from documents and databases), and (4) cadaster (i.e., collected information via previous research or studies).

5 Results

Although, this research presents a series of intermediate results, the methodological proposal to measure the sociocultural-carrying capacity of Rapa Nui is presented in Table 1. The indicator matrix is constructed under six variables. The first corresponds to the "vitality of cultural-artistic expressions," it seeks to elucidate both the participation of the community on community celebrations and commemorations and artistic-cultural expressions, such as music, dance, traditional medicine, *kai kai* (i.e., a traditional game that combines poetry with the realization of figures

Table 1 Methodological proposal to measure the impact of population growth in Rapa Nui

Variable	Secondary variable	Tertiary variable	Indicator and instrument[a]
Vitality of cultural artistic expressions	Celebrations and commemorations	Community involvement	% of Rapa Nui/non-Rapa Nui people who attend community celebrations and commemorations; CCS
		Context of participation	% of Rapa Nui/non-Rapa Nui people who observe or participate in ceremonies; CCS
	Participation in artistic–cultural expressions	Practice	No. of Rapa Nui/non-Rapa Nui people who perform cultural artistic expressions; CCS
		Development of the practice	Level of knowledge acquired by participants from cultural artistic expressions; CCS
		Context of the practice	% of Rapa Nui/non-Rapa Nui people who perform cultural artistic expression in spiritual religious, tourism, and recreational contexts, among others; CCS
	Knowledge carriers	Cultural promoters	No. of cultural promoters by cultural artistic expression; SS and IK
		Dissemination of the practice	No. of people who carry knowledge that disseminates (teaches) the practice; Cad
		Programs of cultural education	% of teaching hours dedicated to cultural artistic education with respect to total teaching hours; IS

(continued)

Table 1 (continued)

Variable	Secondary variable	Tertiary variable	Indicator and instrument[a]
Traditional activities	Participation in some traditional activity (i.e., traditional medicine, agriculture, livestock, fishing, *hahaki*, and gastronomy)	Practice	% of Rapa Nui/non-Rapa Nui people who perform some traditional activity; CCS
	Desertion	Practice	% of Rapa Nui people who have moved from traditional activities to tourism activities; CCS
Vitality of the language	Intergenerational transmission of the vernacular language		% of Rapa Nui/non-Rapa Nui people who speak the vernacular language by age range; CCS
			% of parents who speak to their children in Rapa Nui; CCS
	Proportion of speakers in the population as a whole		% of speakers of the language over the total number of residents of the island; CCS
	Change in the spaces of language use		% of Rapa Nui people who speak the language in public/private spaces; CCS
	Context for learning and teaching the Rapa Nui language		Compliance with the immersion program of the language (compare the established number of hours with the actual number of hours that are being applied); IS

(continued)

Table 1 (continued)

Variable	Secondary variable	Tertiary variable	Indicator and instrument[a]
			No. of professors who speak Rapa Nui (during class hours) over the total number of professors; IS
			% of teaching hours devoted to promoting multilingualism with respect to the total number of teaching hours; IS
	Mastery of the language		Degree of linguistic competence by age range; SS
	Response to new areas		Presence in new areas and the media; Cad
	Willingness to learn		% of Rapa Nui/non-Rapa Nui people who do not speak the language but express an interest in learning it; CCS
Natural cultural heritage	Species for traditional medicine	Preservation status	Level of preservation status; SS
		Threats	Degree of exploitation; Cad
	Species for crafts, sculpture and carving	Preservation status	Level of preservation status; SS
		Threats	Degree of exploitation; Cad
	Species for food	Preservation status	Level of preservation status; SS
		Threats	Degree of exploitation; Cad
	Species for furniture making and construction	Preservation status	Level of preservation status; SS
		Threats	Degree of exploitation; Cad
	Site management		Degree of protection of sites with higher concentration of species; IK

(continued)

Table 1 (continued)

Variable	Secondary variable	Tertiary variable	Indicator and instrument[a]
Archaeological heritage	Threats from land uses	Changes in land use	Distance to the urban expansion radius; Cad
		Livestock uses	Use of the property division in which the site is located; Cad
			Signs of animal transit on the site; Cad
		Agricultural uses	Use of the property division in which the site is located; Cad
		Natural area	Type of predominant vegetation; SS
		Urban area	Distance to the urban area; Cad
			Distance to roads; Cad
	Legal threats		Legal status of the site (public or private); SS
	Population threats	Tourist population	Total tourist population; SS
		Resident population	Total resident population; SS
	Natural threats	Weather conditions	Exposure level; Cad
			Signs of ponding; Cad
		Geographic conditions	Distance to the coastline (m^2); Cad
			Distance to gorges; Cad
			% of water channels on the site; Cad
	Management measures	Compliance with site management measures	Compliance with the site's carrying capacity; Cad
			Compliance with the site's management plan; Cad

(continued)

Table 1 (continued)

Variable	Secondary variable	Tertiary variable	Indicator and instrument[a]
			% of the heritage site under management measures; SS
		Visiting conditions	Presence of fence or entry line (immediate protection); Cad
			Presence of fence or entry line (farther protection); Cad
			Presence of watchmen; Cad
			Presence of visitation trails managed by agents who are in charge of the site's preservation; Cad
			Presence of visitation trails not managed by agencies in charge of the site's preservation; Cad
			Presence of signaling (near protection); Cad
			Presence of signaling (farther protection); Cad
		Heritage education	Presence of heritage education programs; SS
			% of the population who participate in heritage education programs; SS
	Landscape's preservation status	Intervention in the environment	Landscape value (scale used by Sonia Haoa); Cad
	Preservation status of the site	Features	No. of elements of archaeological value on the site; Cad
			Magnitude of grooves; Cad

(continued)

Table 1 (continued)

Variable	Secondary variable	Tertiary variable	Indicator and instrument[a]
			Concentration of elements on the site; Cad
		Signs of damage by anthropic action	Signs of fire; Cad
			Damage caused by traffic or vehicle accidents; Cad
			Signs of re-marking by non-specialist human action; Cad
			% of physical loss or removal; Cad
			Signs of traffic on the archaeological heritage site; SS
		Signs of damage by natural action	Signs of silica crust; Cad
			% of fissures or cracks due to weathering; Cad
			% of natural depression in lithology; Cad
			% of physical loss or removal due to weathering; Cad
			Signs of microflora; Cad
			Signs of corrosion; Cad
			Soil erosion; SS
Coexistence	Identity	Effects of population growth	% of the population that identifies that population growth contributes/does not contribute to the loss of traditional life patterns; CCS

(continued)

Table 1 (continued)

Variable	Secondary variable	Tertiary variable	Indicator and instrument[a]
		Effects of tourism	% of the population that identifies that tourism encourages/does not encourage traditions to be practiced again; CCS
			% of the population that identifies that tourism helps/does not help value the cultural heritage by residents; CCS
			% of the population that identifies that tourism contributes/does not contribute to the loss of traditional life patterns; CCS
	Social conflicts	Effects of population growth	% of the population that identifies that population growth contributes/does not contribute to the development of social conflicts between the Rapa Nui people; CCS
			% of population that identifies that population growth generates/does not generate discrimination; CCS
			% of the population that identifies that population growth contributes/does not contribute to the development of social conflicts between Rapa Nui and non-Rapa Nui people; CCS

(continued)

Table 1 (continued)

Variable	Secondary variable	Tertiary variable	Indicator and instrument[a]
			% of the population that identifies that population growth contributes/does not contribute to the increase of crime, drugs and alcohol; CCS
		Effects of tourism	% of the population that identifies that tourism contributes/does not contribute to the development of social conflicts between residents and tourists; CCS
			% of the population that identifies that tourism contributes/does not contribute to the increase in crime, drugs and alcohol; CCS
	Economic benefits/negative impacts	Effects of population growth	% of the population that identifies that population growth generates/does not generate more employment opportunities; CCS
			% of the population that identifies that population growth increases/does not increase the cost of living; CCS
		Effects of tourism	% of the population that identifies that tourism generates/does not generate greater employment opportunities; CCS

(continued)

Table 1 (continued)

Variable	Secondary variable	Tertiary variable	Indicator and instrument[a]
			% of the population that identifies that tourism increases/does not increase the cost of living; CCS
	Overcrowding and urban density	Effects of population growth	% of the population that identifies that population growth contributes/does not contribute to excessive housing construction; CCS
			% of the population that identifies that population growth contributes/does not contribute to the increase of vehicular traffic and accidents; CCS
		Effects of tourism	% of the population that identifies that tourism contributes/does not contribute to the excessive construction of housing; CCS
			% of the population that identifies that tourism contributes/does not contribute to the increase of vehicular traffic and accidents; CCS
	Effects to the environment	Effects of population growth	% of the population that identifies that population growth generates/does not generate a greater accumulation of waste and garbage; CCS

(continued)

Table 1 (continued)

Variable	Secondary variable	Tertiary variable	Indicator and instrument[a]
			% of the population that identifies that population growth generates/does not generate water and soil pollution and the loss of biodiversity; CCS
		Effects of tourism	% of the population that identifies that tourism generates/does not generate increased accumulation of waste and garbage; CCS
			% of the population that identifies that tourism generates/does not generate water and soil pollution, and the loss of biodiversity; CCS

[a]*CCS* Carrying capacity survey; *IS* Interview at schools; *Cad* Cadaster; *SS* Secondary source; *IK* Interview with key figures

with thread), body art, myths and legends, crafts, and the dissemination of them. This participation can be, in the capacity of an observer and participant, the role of artistic-cultural expression and how non-ethnic residents are involved in community activities and the like. In turn, this variable allows us to see participation of the community's cultural-artistic expression, context of their practice, and level of knowledge of it. It also considers the role of cultural promoters as agents that carry knowledge, transmittable to future generations.

The second revolves around "traditional activities." This variable seeks to deepen the participation of the Rapa Nui community in traditional activities determined by agriculture, livestock, fishing, traditional medicine, *Hahaki* (i.e., seafood), and typical cuisine. It also observes the abandonment of certain traditional activities (e.g., livestock, agriculture, and fishing) from activities related to tourism. The third refers to the "vitality of the language," which is made up of six secondary variables that the United Nations Educational, Scientific and Cultural Organization [45] and Tsunoda [46] indicate are key issues to measure the vitality of the language, including: intergenerational transmission, proportion of speakers in the population as a whole, changes in the fields of use (i.e., if the language is spoken in both public and private space), learning context (i.e., associated with the role of educational institutions), language proficiency, and response to new areas (i.e., presence in the local media). It also includes a variable, indicated as relevant by local experts, that is, motivation to learn the native language. The fourth variable "natural cultural heritage" is defined based on the state of conservation and the threats that affect the natural heritage that is used for traditional medicine; craftsmanship, sculpture, and carving; feeding; furniture; and construction. It also considers the management of tourist sites that have a high concentration of species, such as wetlands. The fifth variable corresponds to "archaeological heritage" and is established from a series of indicators that seek to measure the state of conservation and risk of damage to heritage, either from threats due to land use, legal status, and existing management measures. The sixth variable corresponds to coexistence, which seeks to collect the perception of residents, based on the impact generated by population growth and increase in tourism on different aspects of local daily life.

The measurement matrix and the research process allowed us to build a conceptual definition of sociocultural-carrying capacity in Rapa Nui, which is defined as the territory's capacity to preserve the state of balance and vitality of its cultural and social environment—made up of natural and archaeological heritage, cultural-artistic expressions, traditional activities, orality, and sense of community—all based on the interaction with external figures. This balance can be achieved from two axes, by establishing: limits of demographic-carrying capacity and management instruments in accordance with the sociocultural protection of the territory. In this sense, territory should not exceed sociocultural-carrying capacity by balancing: (1) its identity, understood as the combination between the vitality of its cultural-artistic expressions, traditional activities, and language; (2) a good state of preservation and low risk of damage to its natural, cultural, and archaeological heritage; and (3) a positive coexistence between the different cultures that circulate in the territory. To achieve a balanced state, the conservation of Rapa Nui's heritage and social coexistence incorporate an understanding of internal and external factors interfering with it.

Among some of the internal factors, it can be noted that the Rapa Nui's natural, cultural heritage is closely related with its identity, insofar as intervening or exploiting these resources interfere in traditional medicine, crafts, sculpture, and carving. Also, the deterioration of archaeological heritage causes conflicts in coexistence, to the extent that they are responsible for it. In addition, the development of traditional activities, such as livestock and agriculture affect the state of conservation of archaeological heritage, since these activities require the use of soil and adequate protection. On the other hand, external factors such as the polluting of natural resources, excessive production of garbage, and increase in vehicle fleet—effects associated with high, accelerated population growth—interfere with the coexistence and modes of local life. In the same way, scarce planning and implementation of management measures to protect the heritage may affect conservation efforts and status. As a result, for example, loss of identity can cause external factors to occur through the repercussion in the type of developed tourism (i.e., of special interests), interfering in local economic development. To understand the magnitude of this, it is necessary to emphasize that tourism in Rapa Nui has been the protagonist of its economic development, and the positive consequences have been inescapable, especially considering the limited possibilities of economic diversification, due to the essential obstacles that support its condition due to its isolation. Therefore, reverberated in the developed tourism model, it is evident cultural authenticity of the destination, which can trigger both "touristification" and a vicious circle that dismantles the base of the local economy, can have lasting effects on the quality of local life.

6 Conclusion

The developed research proposes the measurement of the sociocultural-carrying capacity of Rapa Nui; however, indirectly it raises the discussion of what has worked under this concept (i.e., from the literature). In the theoretical framework, the definition coined by López and López [26] present the "maximum level of tourist use, which allows preserving the state of balance of the natural environment of a tourist site, which is composed especially of traditions and customs and historical heritage." In turn, it was identified that only two documents addressed the cultural relevance to measuring the carrying capacity of a territory, which contained a superficial and concise analysis and consideration. In fact, Morillo [47] identifies three types of carrying capacity: (1) ecological (i.e., intensity of use, number of users, or level of environmental degradation acceptable), (2) landscape (i.e., ability to absorb the presence of visitors through a landscape), and (3) perceptual- or social-carrying capacity (i.e., limit of psychological tolerance to the presence of visitors by both residents and tourists). Apart from this example, there are very a few others that reference sociocultural inclusion in the concept. As noted, the arguments for maintaining a balanced sociocultural identity within Rapa Nui involves a deep consideration for its heritage and social coexistence, both at an internal and external levels.

Moreover, the methodology is based on quantitative indicators. Quantitative indicators that have the strength of being less ambiguous but also a series of associated problems, such as high-cost and time-collecting data. Such an approach can present enormous difficulties to establish standards and thresholds, compared to less-complicated qualitative ones. The contributive use of these types of indicators lies in the fact that statistical data can be obtained disaggregated, comparable, and measurable over time. This is a crucial advantage which corresponds to the ability to implement a management model that allows for an evaluation progress and setback in its implementation. As such, it establishes clear and concise goals that can be assessed accurately, that is, policies and strategies pointing toward a clear objective. Yet, it is important to emphasize that this proposed method is not intended to be cross-compatible with other destinations, but to clearly support public policy and the framing development of the new designated government bill. At the same time, it is important to point out that this methodology is not capable of collecting profound aspects of local identity, as seen in the development of this chapter, in which the variable "belief and religion" is left out, due to the methodological difficulty of finding how it may impact demographic growth and tourist consolidation.

Therefore, it should be clarified that the proposed approach has certain limitations. Nonetheless, the benefit is inescapable, from the lifting of this information—by way of example—it is possible to monitor the intergenerational transmission of language, participation of the community in cultural, artistic manifestations, and thus see what behavioral actions over time reflect upon the target. In conclusion, this methodology not only defines the sociocultural-carrying capacity of Rapa Nui but also allows for the implementation of measures in those areas of conflict as a valuable model that promotes sound, management territory-wide.

References

1. Instituto Nacional de Estadísticas (2002) Censo de Población y Vivienda. INE
2. Instituto Nacional de Estadísticas (2017) Pre-censo de Población y Vivienda. INE
3. Biblioteca del Congreso Nacional (2007) Historia de la Ley No 20.573. Reforma Constitucional sobre territorios especiales de Isla de Pascua y Archipielago Juan Fernández. Santiago, Chile
4. Subsecretaria de Desarrollo Regional y Administrativo (2016) Decreto No. 833. Gobierno de Chile, SUBDERE: Informe final consulta indígena: anteproyecto de ley de residencia, permanencia, traslado desde y hacia el territorio especial de Isla de Pascua
5. Tarlombani da Silveira MA (2009) Turismo y sustentabilidad. Entre el discurso y la acción. Urbano 12:61–75
6. Ko DW, Stewart WP (2002) A structural equation model of residents' attitudes for tourism development. Tour Manag 23:521–530. https://doi.org/10.1016/S0261-5177(02)00006-7
7. Rainforest Alliance (2008) Buenas Prácticas para turismo sostenible. Rainforest Alliance, New York
8. Maldonado C (2006) Turismo y comunidades indígenas: Impactos, pautas para autoevaluación y códigos de conducta. In: OIT SEED Work. Pap. No. 79. https://www.ilo.org/empent/Publications/WCMS_117521/lang–es/index.htm. Accessed 3 May 2019
9. Navarro Jurado E, Tejada Tejada M, Almeida García F et al (2012) Carrying capacity assessment for tourist destinations. Methodology for the creation of synthetic indicators applied in a coastal area. Tour Manag 33:1337–1346. https://doi.org/10.1016/J.TOURMAN.2011.12.017
10. UNESCO (2004) What is intangible cultural heritage? Intangible heritage. In: UNESCO. https://ich.unesco.org/en/what-is-intangible-heritage-00003. Accessed 3 May 2019
11. FEDETUR (2012) Barómetro chileno del turismo. Federación de Empresas de Turismo de Chile, Santiago
12. SERNATUR (2013) Seis maneras de disfrutar Isla de Pascua. Rev Q 2:8–14
13. Calderon C, O´Ryan P (2011) Desafíos y oportunidades de desarrollo sostenible de Isla de Pascua basada en el turismo. Santiago, Chile
14. Pérez AS, Mesanat GG, Rozo E (2007) Comparativa de indicadores de sostenibilidad para destinos desarrollados, en desarrollo y con poblaciones vulnerables. Ann Tour Res en Español 9:150–176
15. Uysal M, Sirgy MJ, Woo E, Kim H (Lina) (2016) Quality of life and well-being research in tourism. Tour Manag 53:244–261. https://doi.org/10.1016/J.TOURMAN.2015.07.013
16. Ángel P (2014) Indicadores socioculturales para monitorear el impacto del ecoturismo originado por la existencia el área protegida sobre el medio social. Tesis para optar al título de Geógrafa. Universidad de Chile, Santiago, Santiago, Chile
17. Azevedo L (2007) Ecoturísmo Indígena. Abya-Yala, Quito, Ecuador
18. Konan DE (2011) Limits to growth: tourism and regional labor migration. Econ Model 28:473–481. https://doi.org/10.1016/J.ECONMOD.2010.08.001
19. Leff E (1998) Saber ambiental: Sustentabilidad, racionalidad, conplejidad, poder, 3rd ed. PNUMA, United Nations Environment Programme, Mexico City

20. CNCA (2012) Estudio diagnóstico del desarrollo cultural del pueblo Rapa Nui. Recuperado el 27 de Abril de 2016, de Consejo Nacional de la Cultura y las Artes
21. Blázquez J (2012) Capítulo 2: Impactos, riesgos y limitaciones de los modelos turísticos convencionales: nivel macro-socioeconómico, nivel micro-socioeconómico, medioambiental y sociocultural. In: Rivera M, Rodriguez L (eds) Turismo responsable, sostenibilidad y desarrollo local comunitario. Cátedra intercultural, Córdoba, Spain, pp 43–64
22. Hossain MA, Piyatida P, da Silva Ja T, Fujita M (2012) Molecular mechanism of heavy metal toxicity and tolerance in plants: central role of glutathione in detoxification of reactive oxygen species and methylglyoxal and in heavy metal chelation. J Bot 1–37. https://doi.org/10.1155/2012/872875
23. García Hernández M (2003) Turismo y conjuntos monumentales : capacidad de acogida turística y gestión de flujos de visitantes. Tirant lo Blanch
24. Gabrielsen P, Bosch P (2003) Environmental indicators: typology and overview—European Environment Agency. In: EEA international working paper. https://www.eea.europa.eu/publications/TEC25. Accessed 28 Nov 2018
25. Coccossis H (2017) Sustainable tourism and carrying capacity: a new context. In: Coccossis H, Mexa A (eds) The challenge of tourism carrying capacity assessment: theory and practice. Routledge, pp 19–30
26. López-Bonilla LM, López-Bonilla JM (2005) El periplo sustentable. Universidad de Sevilla, Spain
27. Palazuelo FP (2007) Sostenibilidad y turismo, una simbiosis imprescindible. Estud Turísticos 172:13–62
28. Bretlaender D, Toth P (2014) Kwanini carrying capacity assessment investors government guests Kwanini people workforce
29. CONAF (2017) Estadística visitantes unidad SNASPE. http://www.conaf.cl/wp-content/files_mf/1522175651Totalvisitantes2017.pdf. Accessed 13 May 2019
30. CONAF (2007) Estadística visitantes unidad SNASPE. http://www.conaf.cl/wp-content/files_mf/1385733277Totalvisitantesaño2007.pdf. Accessed 13 May 2019
31. Rendón M Desarrollo Sostenible y Turismo. Modulo, Instituto Latinoamericano de Ciencias, Diplomado gestión del turismo sostenible
32. Poggi M, Ferreira M, Van de Meene D (2015) Competitividad y turismo sustentable. Caso de aguas de San Pedro (San Pablo, Brazil). Estud y Perspect en Tur 220–235
33. Stronza A, Gordillo J (2008) Community views of ecotourism. Ann Tour Res 35:448–468. https://doi.org/10.1016/J.ANNALS.2008.01.002
34. Mai T, Smith C (2015) Addressing the threats to tourism sustainability using systems thinking: a case study of Cat Ba Island, Vietnam. J Sustain Tour 23:1504–1528. https://doi.org/10.1080/09669582.2015.1045514
35. Santos JLQ (2004) Los impactos económicos, socioculturales y medioambientales del turismo y sus vínculos con el turismo sostenible. An del Mus América 263–274
36. Cordeiro ID e, Körössy N, Selva VSF (2013) Capacidade de carga recreativa para embarcações: o caso da área de proteção ambiental de Guadalupe. TURyDES, Tur y Desarro Local 6
37. Dias I, Körössy N, Fragoso V (2012) Determinación de la capacidad de carga turística El caso de Playa de Tamandaré, Pernambuco, Brazil. Estud Perspect tur 21:1630–1645
38. Papageorgiou K, Brotherton I (1999) A management planning framework based on ecological, perceptual and economic carrying capacity: the case study of Vikos-Aoos National Park, Greece. J Environ Manage 56:271–284. https://doi.org/10.1006/JEMA.1999.0285
39. Byron C, Link J, Costa-Pierce B, Bengtson D (2011) Calculating ecological carrying capacity of shellfish aquaculture using mass-balance modeling: Narragansett Bay, Rhode Island. Ecol Model 222:1743–1755. https://doi.org/10.1016/J.ECOLMODEL.2011.03.010
40. Shi H, Shen C, Zheng W et al (2016) A model to assess fundamental and realized carrying capacities of island ecosystem: a case study in the southern Miaodao Archipelago of China. Acta Oceanol Sin 35:56–67. https://doi.org/10.1007/s13131-016-0808-y
41. Gonzalez-Alvarez Y, Keeler AG, Mullen JD (2006) Farm-level irrigation and the marginal cost of water use: evidence from Georgia. J Environ Manage 80:311–317

42. OMT (1998) Introducción al turismo. Madrid, Spain
43. Vereczi G (2007) Sustainability indicators for ecotourism destinations and operations. In: Black R, Crabtree A (eds) Quality assurance and certification in ecotourism. CAB International, Oxford, pp 101–115
44. Organización de las Naciones Unidas para la Agricultura y la Alimentación Rome. Directrices para la recopilación sistemática de datos relativos a la pesca de captura
45. UNESCO (2003) Convention for the safeguarding of the intangible cultural heritage. United Nations Educational, Scientific and Cultural Organization, Paris
46. Tsunoda T (2006) Language endangerment and language revitalization. De Gruyter, Berlin
47. Morillo M (2002) Sustentabilidad socio-ambiental de la actividad turística receptora. Rev Econ 17–18:69–107

Territorial Integration of Foreigners: Social Sustainability of Host Societies

Federico Benassi and Alessia Naccarato

Abstract The foreign population has become a structural trait of Italian society, and its territorial integration a key factor in social sustainability. Mixed couples, an emerging phenomenon in the Italian context, are leading to a change in social space and residential geography of the local environment. Encouraging an improvement in the level of territorial integration of the foreign population, by the host, plays an important role in understanding demographic changes. This chapter proposes a theoretical reflection on the importance of territorial integration of foreigners and an assessment of the dimensional effects it presents to the host societies' social cohesion, and an empirical application to examine the relationships between foreigner residential integration and mixed couples. Results indicate that the increase in mixed couples leads to a weakening of residential segregation and therefore to greater territorial integration of the foreign population in the host society. Taking level of segregation of other ethnic groups under control, the effect of mixed couples on the level of residential segregation remains negative, while at the same time, the level of residential segregation of a given foreign community is positively correlated with other foreign community's level of residential segregation. The territorial integration of foreigners is, thus, strongly linked to the local, social environment in context of multi-segregation. From this perspective, the growth of mixed-race couples, at least within a territorial dimension, can represent an agent of change in the social space modifying the majority and minority groups' residential geography.

Keywords Residential segregation · Mixed couples · Foreign population · Index of dissimilarity · Italy

F. Benassi (✉)
Italian National Institute of Statistics, Rome, Italy
e-mail: benassi@istat.it

A. Naccarato
Department of Economics, Roma Tre University, Rome, Italy
e-mail: alessia.naccarato@uniroma3.it

© Springer Nature Singapore Pte Ltd. 2020
G. T. Cirella (ed.), *Sustainable Human–Nature Relations*,
Advances in 21st Century Human Settlements,
https://doi.org/10.1007/978-981-15-3049-4_3

49

1 Introduction

Italy, a country with a long history of emigration, has now become a country of immigration. In recent decades, due to geopolitical changes, the foreign population in the country has increased intensively, showing clear trends toward a stable settlement. Latest available data provided by the Italian National Institute of Statistics (Istat) records that at the beginning of 2018, more than 5 million foreign citizens, 8.5% of the total population, resided in Italy. The value of this indicator is particularly relevant if we consider in 1991, the ratio was only 0.6%. Utilizing this statistical peculiarity, the issue of immigration, foreign presence, and integration have assumed a central role in the Italian political and public debate. In regard to foreigners, particular attention has been paid to territorial and urban dimensions relating to crime, deprivation, marginality, and security.

The increase of foreign presence in Italy, exacerbated by the media, has given rise to emotional polarization of opposing positions that oscillate between the myth of the "good savage," the defense of an original culture, and political management to handle the current emergency migration crisis. As such, vision of society and integration of foreigners in Italy is still unclear. Traditional models, which have been adopted in European countries with the oldest immigrant history (i.e., France, England, and Germany), have exhibited significant change in recent years. As of yet, solid alternatives have not emerged, at least from an empirical point of view. The current model of economic development, indissolubly linked with a rise in foreign presence, has accentuated immense weight on the host societies' social cohesion. In classical economics, immigrants have been seen as "useful invaders" due to the fact they have been able to compensate economic mismatches. In recent years, reports have narrated the theme of "dangerous invaders", as foreigners are perceived as co-determinants of the economic crisis and thus a danger to their host societies' social structure.

We propose, on the one hand, a theoretical reflection on the importance of territorial integration of the foreign population and an assessment of the dimensional effects it presents to the host societies' social cohesion, while on the other, an empirical application will examine the relationships between foreigners' residential integration and mixed couples. Do mixed couples—an emerging phenomenon in Italy—boost the territorial integration of foreigners? Are there any differences among foreign groups? What are the implications in terms of social sustainability? These themes, quite surprisingly, are missing from the Italian-scientific debate, especially with regards to the territorial dimension of foreign integration versus the development of ghettos.

A theoretical model for tackling these questions is the adoption of spatial assimilation theory in a bi-directional approach, between the majority population (i.e., Italians) and minority groups (i.e., foreigners) of which mixed couples may represent a proxy variable. In fact, mixed couples can be seen both as a product and driver of the integration process in which if couples split between the majority population and minority groups that live together will modify the social space of a given environment. An empirical analysis within the Tuscany region, Italy, is conducted at the municipality level using sub-municipal enumeration areas as rudimentary units. We calculated two basic statistical indicators for all of the 287 municipalities within the Tuscany region and for the selected population groups: (1) index of dissimilarity and (2) percentage of mixed couples. Using the variable "country of citizenship," both indicators were computed for foreigners as a whole and for selected foreign groups: Africans, Americans, Asians, and Europeans. Statistical data used the 2011 Italian population and housing census since it provides fine-scale, geographical data suitable for this type of analysis. The system of relationships between foreign groups, residential segregation and mixed couples, were detected with two sets of linear regression models. A breakdown of the research is structured in three key parts: (1) description of the theoretical approach in measuring residential segregation and an explanation as to why this process is important in relation to the host societies' social sustainability, (2) index adopted for measuring residential segregation and role of mixed couples, and (3) description of the geographical context of the analysis and empirical results.

2 Territorial Integration of Foreigners and the Host Societies' Social Sustainability

From a conceptual and operative viewpoint, the term "territorial integration of foreigners" can be defined as the absence of residential segregation of foreigners. Utilizing spatial assimilation theory, there is no residential segregation if the minority group (i.e., foreigners) is distributed territorially in a similar way to the majority group (i.e., indigenous population). This theoretical approach originates from work conducted at the ecological school of Chicago in which residential segregation of foreigners is seen as a natural process and function of social status [1–4]. It should be noted that the concept of spatial assimilation is increasingly moving away from its original connotation of describing social phenomena that take place spontaneously and often unintentionally over time between majority and minority ethnic groups. In recovering the concept of assimilation, therefore, in an intransitive sense (i.e., becoming similar), it is necessary to keep in mind that the process is not theoretically one way. That is, the process does not exclude the possibility that even majority groups can make some of the minority groups' characteristics become their own. From this viewpoint, mixed couples, here defined as couples in which one member is not an Italian citizen and the other is, represent an event that linking together two members of different social groups (i.e., majority and minority groups) inevitably modifies the social space of a given environment or, in other words, the social space between us [5]. We are well aware that the spatial assimilation approach, although very widespread and still widely adopted in territorial studies, is certainly not the only approach and is not exempt from criticism. Among the alternative visions is one that is ascribable to the model of ethnic status, according to which residential segregation is the fruit of a strategy assumed by the minority group in order to pre-serve and strengthen its own identity [6] and that of the cultural distance between the ethnic minority and the host society [7].

Of course, the greater or lesser level of residential segregation of foreigners, the minority group, may also depend significantly on the majority group's settle-ment choices, according to the phenomenon known as "white segregation" [8–10]. Although not all scholars agree on the idea that higher levels of residential segrega-tion correspond to lower levels of integration [10–12], it seems more widespread the belief that minority groups' residential segregation determines—least at the macro level—a set of negative effects on the host society. In particular, a high level of res-idential segregation can reinforce social exclusion of certain social groups [13, 14] and thus can trigger processes detrimental to social cohesion [15]. From this point of view, therefore, the effects of residential segregation of foreigners on social sus-tainability of the destination country are clear. The social dimension of sustainability has only recently entered the political and scientific debate, having been initially obscured by environmental and economic pillars of the concept of sustainability [16]. In relatively few years, several works have appeared on the subject focusing on the conceptual definition of social sustainability [17], in which almost all attempts to define it, even operationally, the dimension of social cohesion has a fundamental role [18].

The residential segregation of foreigners and the spatial processes that follow are sometimes missing in theoretical models defining the sub-dimensions of social sus-tainability, but it is our precise idea that residential segregation (i.e., a weak level of foreigner territorial integration in a host society) can undermine social sustainability. If we keep in mind that social sustainability has become central to the contemporary political and cultural debate that many scholars have argued as necessary for deter-mining economic and environmental sustainability, then we realize the centrality of the theme. The links between social sustainability and territorial issues can be found in the fact that social sustainability acquires special importance in urban contexts [19]. Spatial and social realities that are multiplying as increasing urban centers become more populated and act as natural attractors of foreign populations [20]. Thus, the relationship between a foreign presence and residential segregation, and social sustainability of a destination country is central to the social changes taking place and the impact on future scenarios.

3   Methodology: Measuring Residential Segregation and the Role of Mixed Couples

Residential segregation of a given population is a term used to indicate the separation between different population groups in a given environment. In terms of residential segregation of foreigners, most often we refer to a multidimensional process whose depiction requires different indices for each dimension. Massey and Denton [21] were the first scholars to define residential segregation as a multidimensional concept, identifying different dimensions measurable by different indices: evenness, exposure, concentration, centralization, and clustering. Although residential segregation is a multidimensional phenomenon, many authors consider the evenness dimension to be the most relevant, and hence, our focus is on this dimension. Evenness involves the differential distribution of the subject population and, generally speaking, indices of evenness measure a group's over- or under-representation in spatial units of a given environment [22]. The more unevenly a population group is distributed across spatial units, the more segregated it is [23]. The most common index in the evenness dimension is the index of dissimilarity.

As such, the first indicator used in the experiment is the Duncan and Duncan's [24, 25] index of dissimilarity (ID). The index is computed using Istat data from 2011 census for the group of foreigners as a whole and for the four selected foreign groups residing in each municipality of the Tuscan region. ID ranges from zero, indicating complete integration, to one, meaning complete segregation [22]. The reference group is the Italian population, that is to say that each foreign territorial distribution is analyzed in comparison to Italians (i.e., the majority group). It is important to keep in mind that the index of dissimilarity is a global measure, which implies that for the computation of each municipality, there is single territorial distribution of each population group at the sub-municipality level (i.e., enumeration areas). At the end of the computational procedure, we have a matrix of 287 cases (i.e., municipalities of Tuscany region) and five vectors of variables (i.e., index of dissimilarity computed for foreigners as a whole as well as for each selected group). ID was computed using Geo-Segregation Analyzer [26].

Many international studies have dealt with analyzing residential segregation of foreigners and their determinants. Recent studies have appeared in the Italian context too, using different perspectives of analysis and geographic scale [27–30]. However, while the issue of the effects mixed couples have on the level of residential segregation of foreigners has been examined in other countries and contexts [31–33], there is a gap in Italy most likely due to the phenomenon of mixed-race couples and families being very limited in the past. Of late, some interesting studies in Italy have appeared [34, 35] in which sample survey data have been used to look at the increase in mixed couples by interpreting the response data as a symptom of integration versus marginalization of a certain type of migrant in search of becoming integrated.

The second indicator analyzed the percentage of mixed couples by computing the ratio between couples (i.e., married or de facto) in which one component is a foreign national out of the total number of couples per 100. The indicator, as in the case of the index of dissimilarity, was computed for each of the 287 Tuscan municipalities, for foreigners as a whole and for each of the four groups. As noted, the research investigated the effect that the percentage of mixed couples have on the level of residential segregation. Concerning USA, the scientific literature stated a high level of residential segregation leads to low rates of partnering between them [36–39]. However, some scholars have begun to question these effects with studying the inverse relationship and finding that the increase in mixed couples and marriages favor a contraction in the levels of residential segregation of foreigners [31]. To measure the effect that mixed-race couples have on the residential segregation of the selected groups, we used two sets of linear regression models. First, we computed five regression models in which the index of dissimilarity (ID) of each population group is the dependent variable, and the percentage of mixed couples of the corresponding population group is the independent variable. Second, we followed up by computing four regression models in which the index of dissimilarity of a given group is the dependent variable, and the percentage of mixed couples of the corresponding group plus the index of dissimilarity of the others foreign groups are the independent variables. In the first stage, the idea was to evaluate the net effect of mixed couples on the level of segregation. In the second stage, the idea was to evaluate the net effect that the mixed couples and the level of dissimilarity of other foreign groups had on the level of given foreign group's residential segregation.

4   Case Study: Tuscany

A regional case study is presented for the Tuscany region, characterized by a high degree of heterogeneity in demographic, socioeconomic, and territorial terms (Fig. 1a). In this context, it can be considered a very encouraging study area for assessing foreign presence and international migration [40]. The regional territory is composed of ten provinces of which half (i.e., Grosseto, Livorno, Pisa, Lucca, and Massa Carrara) are located on the Tyrrhenian coast, while the rest (i.e., Siena, Arezzo, Florence, Prato, and Pistoia) stretch inland to the Tuscan–Emilian Apennines (Fig. 1b). In this geographical framework, we find important historical and university cities such as Florence, previously the capital of Italy, Pisa, the former maritime republic, and Siena. The polycentric structure of the urban fabric ensures a plurality of urban centers, in addition to those already mentioned, interconnected with each other with different specializations that are based in the remaining provincial capitals. Each territory has its own economic specificity which varies greatly, including sectors of marble and mining, port and logistics, craftsmanship and fabric, goldsmith, and floral nursery. The Tuscany archipelago is also quite important, especially in terms of tourism and for being a destination for particular immigration flows [40]; yet, Tuscany is a region that has been affected for many years by migrants who come not only for work reasons (i.e., economic migration) but also for retirement, as in the case of the large British community [41]. Recently, Tuscany has come to represent a social space in which foreign communities of ancient settlement such as Moroccan, Senegalese, and Polish coexist with communities of more recent arrivals, such as Albanian, Romanian, and Ukrainian. Over the years, the foreign presence in Tuscany, as in all of Italy, has increased significantly.

Foreigners counted in the 1991 census did not reach 30,000 units, representing only 0.5% of the regional resident population; however, in 2011, the date of the last census, more than 320,000 foreign residents, 8.8% of the regional population (i.e., 3.7 million), were surveyed in Tuscany. The foreign population's tendency to grow still remains very intense and sharp, as the latest results indicate foreigners residing

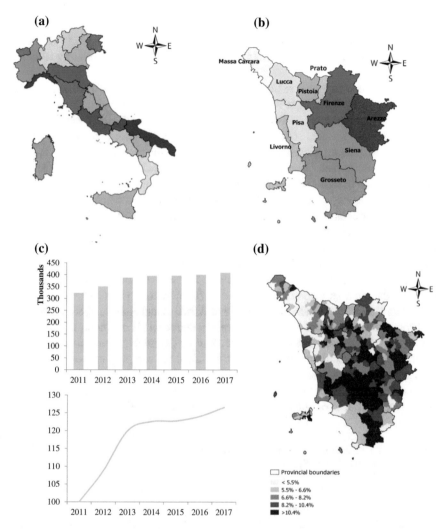

Fig. 1 **a** Italian regions; **b** Tuscan provinces; **c** foreigners residing in Tuscany, absolute number and index number to a fixed base (2011 = 100); **d** foreigners per 100 residents, Tuscan municipalities, 2011 (quantile of the distribution). *Source* **c** authors' elaboration on demographic data provided by Istat; **d** authors' elaboration on 2011 census data

in Tuscany number more than 400,000 (Fig. 1c). From a territorial standpoint, the foreign population is distributed according to specific geographical patterns that underline the importance of local dimensions of the migration phenomenon (Fig. 1d).

In terms of macro-geographical areas, the largest group of foreigners, again in 2011, is represented by Europeans, who make up 58.5% of the foreign population, followed by Asians at 20.4%, Africans at 14.0%, and Americans at 7.1%. One must bear in mind that the aggregation of the country-of-citizenship variable by continents brings together communities that are very different from each other in terms of both demographic structure and migration profile, nonetheless, it was a plausible choice in order to maintain a sufficient number of cases in the variable relating to mixed couples.

As the stable component of the foreign population increased, so did the proportion of mixed couples: 0.6% in 1991, 1.90% in 2001, and 3.0% in 2011. The same indicator at the national level was equal to 0.5%, 1.5%, and 2.4%, respectively. A clear tendency toward a stable settlement of the foreign population is therefore at hand. A very heterogeneous picture emerges in relation to the phenomenon of residential segregation and mixed couples. The index of dissimilarity assumes a high level of variability both in terms of group and in relation to geography. The average level of segregation is quite contained for foreigners considered as a whole (i.e., 0.2950) as well as for Europeans (i.e., 0.3105), Africans fall in an intermediate situation (i.e., 0.5727), while both Americans and Asians record values above 0.6, a limit usually considered to define situations of high residential segregation. The average values of the index of dissimilarity are 0.6050 for the former and 0.6369 for the latter (Table 1 and Fig. 2). A high level of variability is also recorded for the percentage of mixed couples, the average values of which are as follows: foreigners as a whole, 3.3%; Europeans, over 2%; Americans, 0.5%; and Africans and Asians, about 0.2%.

From the values of these indicators we can deduce a set of important evidences. The group of foreigners as a whole shows a low level of segregation, accompanied by a percentage of mixed couples that are still limited but certainly relevant, especially if interpreted from a diachronic perspective.

Relations among individual groups of the foreign population also indicate an important level of variability. The largest group, that of Europeans, is also the most

Table 1 Index of dissimilarity and percentage of mixed couples for foreigners and selected foreign groups, Tuscan municipalities, 2011

	N	Min	Max	Mean	Sd
Index of dissimilarity					
Africans	287	0.1925	0.9822	0.5727	0.1435
Americans	287	0.2105	0.9778	0.6050	0.1441
Asians	287	0.2853	0.9995	0.6369	0.1566
Europeans	287	0.0594	0.6071	0.3105	0.0798
Foreigners	287	0.0870	0.5064	0.2950	0.0704
Percentage of mixed couples					
Africans	287	0.0000	1.5385	0.2307	0.1961
Americans	287	0.0000	2.1429	0.4695	0.3433
Asians	287	0.0000	1.6129	0.1875	0.2066
Europeans	287	0.0000	6.4935	2.2006	0.9237
Foreigners	287	0.0000	8.2305	3.2804	1.1841

Source Author elaboration on 2011 census data

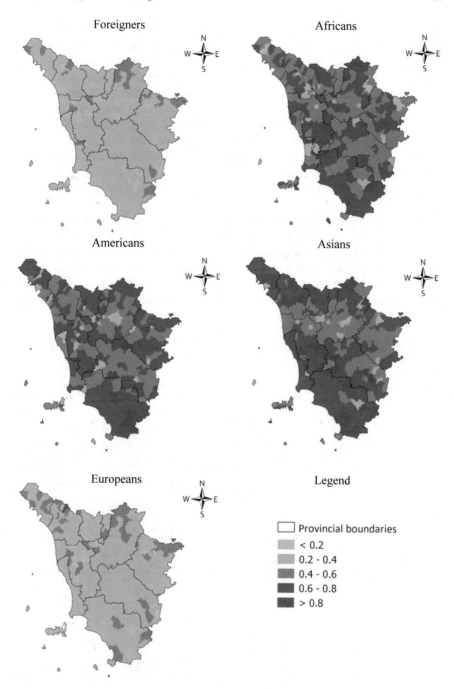

Fig. 2 Index of dissimilarity of foreigners and selected foreign groups, Tuscan municipalities, 2011. *Source* Authors' elaboration on 2011 census data

territorially integrated, with the highest percentage of mixed couples. The approach of cultural distance as well as social status does not seem refutable. In opposite, we find Asians, who despite being the second foreign group in terms of number, have the lowest level of territorial integration (i.e., highest level of segregation and lowest percentage of mixed couples). In this case, the model of cultural distance, as well as that of voluntary segregation, or ethnic status does not seem to be refutable. Next, Africans, the third-largest group in 2011, are the least segregated, after Europeans, even though the percentage of mixed couples is the second lowest after that of Asians. Finally, we find Americans who are the smallest community with little integration from a territorial viewpoint and index of dissimilarity second highest after that of Asians, but which records, after the Europeans, the second-highest percentage of mixed couples.

Results of the regression analysis help us to focus more on what has emerged to date. The first regression models indicate a net effect of mixed couples on the index of dissimilarity as always negative (i.e., as the percentage of mixed couples increases, the level of residential segregation decreases). This effect is detected for foreigners as a whole and for each of the groups (Table 2). Belonging to one ethnic group rather than to another, except in relation to the level of dissimilarity, does not

Table 2 First set of regression models; ID = index of dissimilarity; PMC = percentage of mixed couples

Models	Beta	p-value
ID Africans	–	–
PMC Africans	−0.209	0.000
N (287)		
Adj R^2 (0.044)		
ID Americans	–	–
PMC Americans	−0.379	0.000
N (287)		
Adj R^2 (0.144)		
ID Asians	–	–
PMC Asians	−0.399	0.000
N (287)		
Adj R^2 (0.159)		
ID Europeans	–	–
PMC Europeans	−0.226	0.000
N (287)		
Adj R^2 (0.051)		
ID Foreigners	–	–
PMC Foreigners	−0.149	0.011
N (287)		
Adj R^2 (0.022)		

Source Authors' elaboration on 2011 census data

Table 3 Second set of regression models; ID = index of dissimilarity; PMC = percentage of mixed couples

Models	Beta	p-value	Models	Beta	p-value
ID Africans	–	–	*ID Asians*	–	–
ID Americans	0.162	0.004	*ID Africans*	0.211	0.000
ID Asians	0.223	0.000	*ID Americans*	0.182	0.001
ID Europeans	0.250	0.000	*ID Europeans*	0.150	0.007
PMC Africans	−0.248	0.000	*PMC Asians*	−0.326	0.000
N (287)			*N* (287)		
Adj R^2 (0.261)			Adj R^2 (0.310)		
ID Americans	–	–	*ID Europeans*	–	–
ID Africans	0.226	0.000	*ID Africans*	0.237	0.000
ID Asians	0.247	0.000	*ID Americans*	0.175	0.002
ID Europeans	0.175	0.001	*ID Asians*	0.242	0.000
PMC Americans	−0.442	0.000	*PMC Europeans*	−0.272	0.000
N (287)			*N* (287)		
Adj R^2 (0.372)			Adj R^2 (0.279)		

Source Authors' elaboration on 2011 census data

seem to have a selection effect; the effect is, however, stronger for the population groups with higher levels of dissimilarity.

The results of the second regression models indicate the level of residential segregation of each group as related to the percentage of mixed couples in the reference group plus the level of segregation of the other population groups. This result is very similar to the first stage. However, the idea is to best understand if, and to what extent, a situation of segregation of one group is unfavorable even for the level of segregation of the other group and if, at the same time, the effect exerted by the variable percentage of mixed couples remained stable. The results of the four estimated models are given in Table 3.

Overall, if the levels of residential segregation are equal, the effect of mixed couples on the dependent variable remains negative; moreover, as the percentage of mixed couples increases, on average, the level of territorial integration increases. It is evident that the most intense effect is shown by Americans, followed by Asians, Africans, and finally Europeans. Keeping the effect of residential segregation of other communities under control, the greatest effects do not appear in communities with higher levels of residential segregation. Thus, keeping the effect of mixed-race couples under control, one group's segregation is always a growth factor of the other ethnic group's segregation. This is confirmed in all possible combinations, given the selected four groups.

It seems reasonable to assume local factors affect the groups' segregation levels rather than their belonging to one ethnic group or another. In fact, this societal phenotype can be labeled as multi-group residential segregation; it seems that in municipalities where the level of residential segregation is high for one ethnic group, it is more likely that it is so for the others. There is, therefore, something that can be defined as social space between majority and minority groups that define the level of general residential segregation. However, if we consider the fact that, for all the groups as well as for the foreigners considered as a whole, the results of the first model show an increase in mixed couples that produces a deterrent effect on the level of segregation. This effect can favor the decrease not only of the segregation of a specific group but also that of other groups. Rendering the theoretical approach used in this study, the process of formation of mixed couples is, at least from a territorial standpoint, an event that changes the residential geography of a given environment and thus its social space.

5   Conclusion

The territorial integration of foreigners is a key factor for guaranteeing the host societies' social sustainability. The existence of mixed couples is typically seen as a sign of foreigners integrating [42], but at least in the Italian context, this phenomenon has not yet been deeply examined, especially in terms of its effect on the residential segregation of foreigners. This is partially due to the prevalence of mixed couples in Italy as relatively new. From a theoretical outlook, there are two main interpretations that require particular attention: mixed couples and the integration process. According to the assimilation theory, mixed couples are an effect of the integration process [3]; while, using the exchange theory, mixed couples are a driving force [43]. The achieved findings show that the increase in mixed couples leads to a weakening of the residential segregation and therefore to greater territorial integration of the foreign population in the host society. The effect is stronger for the population groups with higher levels of dissimilarity. This can be interpreted to mean the mixed couples phenomenon which could play a crucial role in the process of migrants' territorial integration. Taking the level of segregation of other ethnic groups under control, the effect of mixed couples on the level of residential segregation remains negative, and at the same time, the level of residential segregation of a given ethnic group is positively correlated with other ethnic group's level of residential segregation.

The territorial integration of foreigners is, therefore, strongly linked to the local, social environment in context of multi-segregation. From this perspective, the growth of mixed couples, at least in a territorial dimension, can represent an agent of change in the social space modifying the majority and minority groups' residential geography. From this point of view, the social impact and consequences of mixed couples are absolutely relevant and require a reflection on future population scenarios and research topics to be followed. More in-depth studies are clearly needed, especially since residential segregation is a phenomenon that requires diachronic observations over time. As such, census data in Italy from mixed couples, due to marginal numbers, is limited to 2011. The next demographic census will add knowledge to a much-needed topic insufficiently investigated Italy-wide.

References

1. Zorbaugh HW (1983) The gold coast and the slum: a sociological study of Chicago's near Northside. University of Chicago Press, Chicago
2. Whyte WF (1993) Street corner society: the social structure of an Italian slum. University of Chicago Press, Chicago
3. Gordon MM (1964) Assimilation in American life: the role of race, religion, and national origins. Oxford University Press, Oxford
4. Park RE, Burgess EW, McKenzie RD (1925) The city. University of Chicago Press, Chicago
5. Enos RD (2017) The space between us: social geography and politics. Cambridge University Press, Cambridge
6. Simpson L, Finney N (2010) Parallel lives and ghettos in Britain: facts or myths? Geography 95:124–131
7. Boal FW (2013) Exclusion and inclusion: segregation and deprivation in Belfast. In: Mustered S, Ostendorf W (eds) Urban segregation and the welfare state inequality and exclusion in Western cities. Routledge, London, pp 94–109
8. Phillips D (2006) Parallel lives? Challenging discourses of British Muslim self-segregation. Environ Plan D Soc Sp 24:25–40. https://doi.org/10.1068/d60j
9. Bolt G, van Kempen R, van Ham M (2008) Minority ethnic groups in the Dutch housing market: spatial segregation, relocation dynamics and housing policy. Urban Stud 45:1359–1384. https://doi.org/10.1177/0042098008090678
10. Bolt G, Özüekren AS, Phillips D (2010) Linking integration and residential segregation. J Ethn Migr Stud 36:169–186. https://doi.org/10.1080/13691830903387238
11. Barth F (1969) Ethnic groups and boundaries: the social organization of culture difference. The Little, Brown and Company, Boston
12. Musterd S (2003) Segregation and integration: a contested relationship. J Ethn Migr Stud 29:623–641. https://doi.org/10.1080/1369183032000123422
13. Williams DR, Collins C (2001) Racial residential segregation: a fundamental cause of racial disparities in health. Public Health Rep 116:404–416. https://doi.org/10.1093/phr/116.5.404
14. Nijkamp P, Poot J (2015) Cultural diversity: a matter of measurement. Discuss Pap Ser Inst Study Labor No 8782:1–40
15. van Ham M, Manley D (2009) Social housing allocation, choice and neighbourhood ethnic mix in England. J Hous Built Environ 24:407–422. https://doi.org/10.1007/s10901-009-9158-9
16. Colantonio A (2009) Social sustainability: a review and critique of traditional versus emerging themes and assessment methods Book section Social sustainability: a review and critique of traditional versus emerging themes and assessment methods. In: SUE-Mot conference 2009: second international conference on whole life urban sustainability and its assessment: conference proceedings, pp 865–885
17. Eizenberg E, Jabareen Y (2017) Social sustainability: a new conceptual framework. Sustainability 9:68. https://doi.org/10.3390/su9010068
18. Vallance S, Perkins HC, Dixon JE (2011) What is social sustainability? A clarification of concepts. Geoforum 42:342–348. https://doi.org/10.1016/J.GEOFORUM.2011.01.002
19. James P (2015) Urban sustainability in theory and practice: circles of sustainability. Routledge, London
20. Strozza S, Benassi F, Ferrara R, Gallo G (2016) Recent demographic trends in the major Italian urban agglomerations: the role of foreigners. Spat Demogr 4:39–70. https://doi.org/10.1007/s40980-015-0012-2
21. Massey DS, Denton NA (1988) The dimensions of residential segregation. Soc Forces 67:281–315. https://doi.org/10.2307/2579183
22. Iceland J, Weinberg DH, Steinmetz E (2002) Racial and ethnic residential segregation in the US: 1980–2000. US Government Printing Office, Washington

23. Martori JC, Apparicio P (2011) Changes in spatial patterns of the immigrant population of a southern European metropolis: the case of the Barcelona metropolitan area (2001–2008). Tijdschr voor Econ en Soc Geogr 102:562–581. https://doi.org/10.1111/j.1467-9663.2011. 00658.x
24. Duncan OD, Duncan B (1955) A methodological analysis of segregation indexes. Am Sociol Rev 20:210–217. https://doi.org/10.2307/2088328
25. Duncan OD, Duncan B (1955) Residential distribution and occupational stratification. Am J Sociol 60:493–503. https://doi.org/10.1086/221609
26. Apparicio P, Forurnier E, Apparicio D (2019) Geo-segregation analyzer: a multi-platform application (Version 1.1). INRS Urbanisation Cult Soc: Spat Anal Reg Econ Lab
27. Cristaldi F (2002) Multiethnic Rome: toward residential segregation? GeoJournal 58:81–90. https://doi.org/10.1023/B:GEJO.0000010827.68349.9e
28. Busetta A, Mazza A, Stranges M (2015) Residential segregation of foreigners: an analysis of the Italian city of Palermo. Genus 71:177–198. https://doi.org/10.4402/GENUS-688
29. Benassi F, Lipizzi F, Strozza S (2017) Detecting Foreigners' spatial residential patterns in urban contexts: two tales from Italy. Appl Spat Anal Policy 1–19. https://doi.org/10.1007/s12061-017-9243-5
30. Benassi F, Heins F, Lipizzi F, Paluzzi E (2018) Measuring residential segregation of selected foreign groups with aspatial and spatial evenness indices. A case study. Springer, Cham, pp 189–199
31. Ellis M, Holloway SR, Wright R, East M (2007) The effects of mixed-race households on residential segregation. Urban Geogr 28:554–577. https://doi.org/10.2747/0272-3638.28.6.554
32. Holloway SR, Ellis M, Wright R, Hudson M (2005) Partnering 'out' and fitting in: residential segregation and the neighbourhood contexts of mixed-race households. Popul Space Place 11:299–324. https://doi.org/10.1002/psp.378
33. Iceland J, Nelson KA (2010) The residential segregation of mixed-nativity married couples. Demography 47:869–893. https://doi.org/10.1007/BF03213731
34. Azzolini D, Guetto R (2016) La crescita delle unioni miste in Italia: un indicatore di integrazione o di marginalità degli immigrati? In: La società italiana e le grandi crisi economiche 1929–2016, 25–26 November 2016. University of Rome La Sapienza, Rome
35. Guetto R, Azzolini D (2015) An empirical study of status exchange through migrant/native marriages in Italy. J Ethn Migr Stud 41:2149–2172. https://doi.org/10.1080/1369183X.2015. 1037725
36. Abrams RH (1943) Residential propinquity as a factor in marriage selection: fifty year trends in Philadelphia. Am Sociol Rev 8:288–294. https://doi.org/10.2307/2085082
37. Clarke AC (1965) An examination of the operation of residential propinquity as a factor in mate selection. Am Sociol Rev 17:17–22. https://doi.org/10.2307/2088355
38. Morgan BS (1981) A contribution to the debate on homogamy, propinquity, and segregation. J Marriage Fam 43:909–921. https://doi.org/10.2307/351347
39. Kalmijn M, Flap H (2001) Assortative meeting and mating: unintended consequences of organized settings for partner choices. Soc Forces 79:1289–1312. https://doi.org/10.1353/sof.2001. 0044
40. Benassi F, Porciani L (2010) The dual demographic profile of migrants in Tuscany. Demographic aspects of migration. VS Verlag für Sozialwissenschaften, Wiesbaden, pp 209–226
41. King R, Patterson G (1998) Diverse paths: the elderly British in Tuscany. Int J Popul Geogr 4:157–182. https://doi.org/10.1002/(SICI)1099-1220(199806)4:2%3c157:AID-IJPG100%3e3.0.CO;2-G
42. Alba RD, Golden RM (1986) Patterns of ethnic marriage in the united states. Soc Forces 65:202–223. https://doi.org/10.2307/2578943
43. Merton RK (1941) Intermarriage and the social structure. Psychiatry J Study Interpers Process 4:361–374. https://doi.org/10.1080/00332747.1941.11022354

Sustainable Land Reforms and Irregular Migration Management

Samuel W. Mwangi and Giuseppe T. Cirella

Abstract Land-related issues are directly and indirectly the main cause of instability, conflict, and violence in Africa. When we think of addressing the root cause of human displacement and irregular migration in Africa, resolving historical land injustice, mitigating marginalization, and implementation of sustainable land reform ought to be central. Yet, land reform is inadequately considered from the international community toward finding solutions for instability and irregular migration. Land reform legislation hardly shows up in the European Union (EU)-Africa partnership on migration management major policy documents. This research asserts that stability and sustainable migration management in Africa are grounded on enhanced social inclusion established through sustainable land reforms. It also brings to the limelight the disconnection between international relations with Africa in addressing irregular migration and the real threat facing some African communities and households. The linked concept of human security to land security and relational ties to socio-ecological vulnerability and resilience is examined. An exploratory sustainable land reform option is considered as a comprehensive perspective of irregular migration management within the EU-Africa mobility framework.

Keywords Vulnerability · Irregular migration · The EU-Africa partnership · Sustainable land reform

1 Mapping Out the Root Cause

The increasing African irregular migration to Europe demands a reenergized search for the root causes and the means to address them [1–7]. Most of the obvious causes of irregular migration are conflict and violence in many African countries. Even in

S. W. Mwangi (✉)
Institute of Political Science, Tübingen University, Tübingen, Germany
e-mail: sawaamy@gmail.com

G. T. Cirella
Faculty of Economics, University of Gdansk, Sopot, Poland
e-mail: gt.cirella@ug.edu.pl

© Springer Nature Singapore Pte Ltd. 2020
G. T. Cirella (ed.), *Sustainable Human–Nature Relations*,
Advances in 21st Century Human Settlements,
https://doi.org/10.1007/978-981-15-3049-4_4

countries without war or natural disasters responsible for mass displacement, extreme poverty is a key push factor [8–10]. In the effort of managing irregular migration sustainably, the most crucial task is identifying the root causes and locating them in the larger international debate on irregular migration management [11]. Overall, there is a consensus within the literature of African development that land resources and land governance are some of the key factors of the continent's poverty and prosperity [12, 13]. However, little has been done to link the historical land issues as the root causes and irregular migration in Africa—a gap this chapter attempts to fill.

Despite the number of sources showing the need to address these root causes the concept still lacks a universal definition. Addressing the root causes thus remains unclear, especially in the case of irregular migration in Africa. Carling and Talleraas [7] address the causes in a more systematic way as "conditions of states, communities, and individuals that underlie a desire for change, which, in turn, produces migration aspirations." According to Boswell et al. [14], they refer to "underlying structural or systemic conditions which create the pre-conditions for migration or forced displacement." The aspiration of people to break from the chains of structural and systematic poverty is what compels them to actively search for a better life [15]. The dissatisfaction in remaining in their native country increases when the prospect for improved livelihood is slim. This discontent can become even more scorching when people are aware of the existing socioeconomic inequalities but are not capable of changing them through legal or legitimate means. This is the point at which conflict and violence become an alternative means to articulate needs. Moreover, root causes include factors that force people to seek security in the short term while at the same time searching for better living standards in the long term.

Overall, there appears to be a scholarly consensus to reexamine Africa's historical past in order to understand its root causes of the current fragility, conflict, and violence. Despite diverse timelines in investigating the historical past, most scholars on the European Union (EU)-Africa partnership on migration control appear to start their analysis from the engagement in the twentieth century through colonialism [16–24]. We, therefore, define the root causes as the long-standing socioeconomic and political conditions that generate fragility within institutions, vulnerability among people, and violent conflict.

2   Beginning of Land Reform in the Post-colonial Era

The colonial era marked the beginning of modern land-related issues and conflicts [25, 26]. This, however, does not mean that pre-colonial Africa was conflict-free. Definitely, with the rapidly growing population in the continent, idle pieces of land are non-existent. Yet, pressure on land and land resources themselves cannot be the sole cause of conflicts and instability; rather, it is pressure on land resources without proper land governance as one of the key factors [27]. The establishment of new post-colonial independent governments was expected to not only establish structures for future land governance but also rectify marginalization of some communities that already existed during colonialism [28, 29]. This is by acknowledging that the divide-and-rule colonial strategy left many indigenous landowners and communities displaced [30]. Sequently, the modern land governance and legislation came in the wake of the introduction of self-rule in many African countries. This included the establishment of land mapping, land tenure system, landownership and transfer rights, public and private lands, and every other task involved in land governance. The newly independent nations were to start from negative (i.e., correctional level) and not from scratch, bearing in mind the Herculean task that awaited these young nations' effectiveness to redistribute land to displaced persons.

After independence, the weak legal and legislative framework, corruption, and impunity witnessed in most African countries hindered fairness in resolving preexisting land issues [31–33]. Worthwhile noting, the effectiveness of post-colonial land reforms differs from one country to another [34]. Generally, within countries, the commendation of land through post-colonial land reforms initiatives was in favor of the educated and the politically connected. On the other hand, the politically unconnected, the poor, and uneducated remained marginalized. Even among the marginalized populations, some groups were more disregarded than others. For instance, land reforms did not capture the extra vulnerability of women [35–37]. The struggle for fairness among victims of this injustice began. With population increase, the effects of marginalization became more profound, and so did the land-related conflicts and consequent human displacement [38]. Today, these issues have become catastrophic in terms of conflict and civil wars [39, 40]. Violence manifests both at household and community levels, leading to individuals, households, and mass displacements in the current Africa. Chronic disputes between sedentary farmers and nomadic and pastoralist communities, as well as ethnic groups, have forced millions of people out of their homes. The conflict between farmers and pastoralists, especially, in Nigeria, Mali, Niger, and Ethiopia has taken more lives than terrorism [41]. Africa being an agrarian economy, lack of land access has meant a lack of livelihood to most households and, hence, the high prevalence of hunger. Landlessness and marginalization have caused rampant cases of radicalization among youths as well as the emergence and spread of terrorism.

3 Land Injustice and Marginalization as Migration Push Factors

In common parlance, one of the most prevalent categories of conflict in Africa is political violence. Here, political violence is downplayed as a categorization of conflict, but rather, is perceived as the politicization of real socioeconomic concern. In this context, such concern includes land injustice and social and economic marginalization. Analytically, using the conflict tree assessment [42], land injustice (e.g., unfair distribution of land) is a static factor founded at the root. Alternatively, the baseline conflict analysis (i.e., rates and ratios of violence over time of an area) [43] enables for the proper understanding of historical land injustice and communities' marginalization as the root cause of contemporary conflict, including violence and instability. Consequently, these are migration push factors across and outside Africa. To broaden the analytics of this perspective, we need to link irregular migration-specific cases of conflicts over land and marginalization, and trace their origin. As such, the irregular migration at the international level reflects irregular migration at the national and regional levels.

There are numerous cases of conflict-caused displacement rooted in land injustice and marginalization in Africa. In Kenya, what was commonly known as the 2007 post-election violence, and that led to the death of more than a thousand people and displacement of hundreds of thousands, was later diagnosed by the Truth, Justice and Reconciliation Commission of Kenya (TJRC) as a conflict primarily based on historical land issues—right from independence [44–47]. The TJRC report further indicated that the lack of adequate legal and fair channels to express grievances over land injustice has led to the rise and proliferation of radicalization in some regions within the country. "Land-related injustices at the Coast (of Kenya) constitute one of the key reasons for underdevelopment in the area and lie at the root of the emergence of the Mombasa Republican Council (MRC). The emergence of the Sabaot Land Defense Force (SLDF) was due to the government's failure to address land-related injustices that members of the Sabaot have suffered for a long time" [44]. Both MRC and SLDF were rebel groups associated with violence and terrorism.

West Africa is one of the key sources of most irregular migration to Europe. Unsolved land issues and lack of proper channels to address them are the root causes of extreme poverty and radicalization. This has led to the spread of terror groups [48]. Marc et al. [49] state that "drivers of fragility in West Africa are represented in the lack of clarity around landownership, the neglect and the marginalization of peripheral regions, which are also often border regions, and the demographic challenge posed by an increasingly youthful population demanding greater inclusion, in particular through jobs and livelihood opportunities." Vulnerability in Cameroon, for instance, and irregular immigration to Europe present some of the most scorching effects of colonialism. Cameroon was colonized by Germany from 1884 to 1916. After the First World War, under the mandate of the League of Nations, Cameroon was given to Britain and France. Each country introduced its own colonial rule. After independence, post-Cameroon was left divided as French-speaking Cameroon (i.e., Francophone Cameroon), and English-speaking (i.e., Anglophone Cameroon). Today, one the major development challenges is, what is generally referred to as the "Anglophone problem". This problem is not just bilingualism but rather a conflict between two governance systems, including education and law (i.e., British common law *vis-à-vis* French civil law) [50]. Political and socioeconomic marginalization of Anglophone minority Cameroonians is a reality, defined by higher poverty levels as compared to Francophone Cameroon. The struggle for inclusion has led to the escalation of instability and politics of secession [51–53].

The conflict in Mali can be traced back to the 1950s. Currently, Tuareg, who perceive themselves as marginalized and discriminated against, fight for more inclusion

[54]. In Nigeria, the level of development in regions occupied by Hausa Fulani (i.e., pastoralists) is very low compared to regions occupied by other communities [55]. One of the contributing factors for the spread of Boko Haram has been historical, political, and economic marginalization among people in the northeast of Nigeria [56, 57]. The Horn of Africa is another region with the highest forced displacement and irregular migration. Countries such as Somalia, Sudan, southern Sudan, Eritrea, and Ethiopia experience high instability amidst ethnic conflicts which are linked to the competition and struggles over land resources [48]. In the Democratic Republic of Congo, poor governance of mining and minefields has resulted in chronic conflicts. Most African countries have a low institutional capacity and coordination for effective land legislation. Social factors, in particular, lack of proper mechanisms for social inclusion, are the major determinants of vulnerability [58]. The minimal operative social inclusion mechanisms have increased vulnerability to some communities within populations. According to the United Nations International Strategy for Disaster Reduction, vulnerability is viewed as the internal component of risk and generally depicted in terms of exposure and sensitivity [59]. The vulnerability, in this context, involves not only people's subjectivity to external shocks such as drought but also communities' sensitivity to shock [60]. There is a possibility that people are forced to migrate not necessarily because of the magnitude of the external shock, but because of their sensitivity to any disruption. Either way, low resilience is the ultimate cause of irregular (i.e., involuntary) migration. There are various factors that influence vulnerability including environmental challenges, technological constraints, and inadequate structural and management. Environmental hazards are increasingly becoming a migration push factor in Africa where the most impoverished are the most disadvantaged [61, 62]. In Ethiopia, for instance, other than displacement due to conflict, a series of droughts have pushed many people out of their homes [63]. In 2018, almost a million Ethiopians were internally displaced, and a high number migrated to Europe and the Middle East [64].

4　The EU-Africa Partnership on Irregular Migration Management

It is clear that sustainable land reform is one of the solutions to the root causes of displacement and irregular migration in Africa. Even in the EU-Africa partnership framework, the term irregular migrants is a generic term that consists of both people seeking asylum in accordance with asylum law and economic migrants without legal entry documentation. An economic migrant defined by the European Commission is "a person who leaves their country of origin purely for economic reasons that are not in any way related to the refugee definition, in order to seek material improvements in their livelihood" [65]. In other words, irregular migration can be viewed as involuntary migration or forced displacement while in legal terms, it refers to an undocumented or unauthorized entry. The rise of irregular migration to Europe is worrying and calls for intervention. As a result, there has been a series of summits and conventions between African states and the EU and its member states (MSs) in search of a proper approach of irregular migration management. Some of the discussion forums and agreements include: Africa–Europe Summit in Cairo, Egypt, in 2000 [66]; France and African Heads of States Summit in Bamako, Mali, in 2005; Euro-Africa Conference in Rabat, Morocco, in 2006 [67]; Europe-Africa Summit in Lisbon, Portugal, in 2007 [68]; Africa-EU Summit in Tripoli, Libya, in 2010 [69]; EU-Africa Summit in Brussels, Belgium, in 2014 [70]; Valletta Summit in Malta in 2015 [71]; AU-EU Summit in Abidjan, Côte d'Ivoire, in 2017 [72]; and Africa Summit held by the European Commission in Brussels, Belgium, in 2019 [73]. The partnership has also been formally institutionalized through Khartoum Process and Rabat Process. After observing how land and land-related historical injustice and marginalization has created conflict, vulnerability, and displacement, one question we cannot afford to ignore is how has the EU-Africa partnership on irregular migration control articulated the concern of land reform as one of the comprehensive migration management agenda? To effectively answer this, there is a need to analyze major EU-Africa policy documents on irregular migration control and analyze major objectives (i.e., highlighted agendas) by the above summits between the two partners and trace the elements of land reforms.

4.1   Detrimental Effect of the EU-Africa Partnership on Migration Control

Efforts to develop the EU-Africa partnership on migration control have intensified over the last two decades. In 2018, the European Council on Foreign Relations (ECFR) bestowed findings that presented recent insecurities facing Africa as a whole. The ECFR's presentation outlined a common EU-centric standpoint which can be seen, to some degree, as a driver of irregular migration. Up to date findings from the ECFR are illustrated in Fig. 1.

The EU's perspective classifies African challenges into two. On the one hand, there are real issues that are a threat to African people in war-torn and fragile countries. As illustrated, many countries in Africa are experiencing violent conflict and instability. The nature of conflict varies from ethnic conflict to radicalized groups and terrorism. Food insecurity is a major challenge in all countries experiencing conflict. Violent conflict, food insecurity, and extreme poverty are the major drivers of irregular migration. On the other hand, there is the border issue. More than half of African countries experience border disputes. For example, the border between Ethiopia and Eritrea has high levels of volatility and has caused displacement of many people. Other inter-state borders, including North Africa, largely remain non-violent. However, the EU constructs border in the Sahara region (i.e., illustrated as a limited statal border control zone in Fig. 1), as a key threat to African people. Borders are increasingly becoming non-issues to many countries, especially when the African

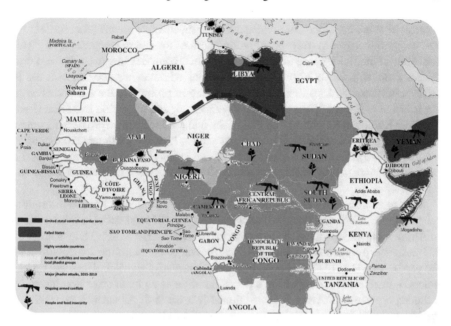

Fig. 1 Map of fragility, conflict, and violence in Sub-Saharan Africa. Adapted from the ECFR [74]

Union is working to open-up the continent to promote the free movement of people and to widen economic prospects. Contrastingly, the EU maintains a different stance of financing border surveillance and control projects that aim to prevent migration to Europe. The construct of the EU's own border management and surveillance capabilities has increasingly become its top priority reflecting the relational direction of the EU-Africa partnership.

In Niger, for instance, a report by the UN Special Rapporteur on the human rights of migrants termed the EU-funded border control program as not only a disruption of people's welfare but also a critical violation of human rights. The report indicates that the EU-Niger partnership has rendered migrants more vulnerable to abuse, exploitation, and violence. The situation is similar in Mali where the EU has had a migration control partnership with neighboring countries. The partnerships have led to the rise of human traffickers and smugglers who exploit migrants in every form. Government agencies, working in collaboration with the EU, conduct arbitrary arrests and detention and expulsion of migrants. In other cases, migrants and asylum seekers lacking access to food, water, and healthcare as well as unaccompanied children and women have been key victims of the partnership agreements on migration control [74]. Globally, in terms of the Human Development Index, Niger is ranked last. This raises the question as to whether the EU has taken advantage of poverty to advance its interests in irregular migration control.

When the African partner countries resisted cooperation toward border protection and surveillance, the EU introduced a "more for more" mechanism migration management partnership [75]. This means more cooperation toward irregular migration

control for more development aid. Equally, the mechanism means "less for less"—
less cooperation toward irregular migration control for less development [76]. This
approach of the EU relating to the African partner countries has strongly been criticized. First, it presents a threat of reducing or withdrawing development aid to
African partner countries that resist cooperating in promoting the EU's version of
the solution to the challenge of forced displacement and migration in Africa via its
Southern European Neighborhood Policy. Second, the mechanism provides incentive
to corrupt regimes and failed African states such as Libya to participate in migration
control [77–79]. The EU's assistance to Africa is thus losing credibility as development aid increasingly becomes remuneration for the work of irregular migration
control.

Such retrogressive external action in Africa has not gone unnoticed in migration-
related studies. Diez and Pace [80] explains that "the EU humanitarian interventions
in Africa seem to depend on the geostrategic interests of EU MSs". Many studies on
EU partnership agreements with the African countries on migration control illustrate
the gap between the EU's self-perception as a 'normative actor' and its view by 'the
others'—in reference to Elgström's [81, 82] notion as a global actor dominated by
a hidden agenda and driven by self-interest [80, 83, 84]. As such, it appears that
the EU will violate human rights, in the name of development assistance, to prevent
migrants from reaching Europe.

5 Land Reform as a Missing Agenda in the EU-Africa Partnership on Migration Management

The aim of highlighting objectives is to demonstrate the absence of land reform
as part of the agenda. A summary of the objectives (i.e., presumed to be a part
of the agenda) of various irregular migration control agreements and action plans
between the EU and African partners is given in Table 1. The majority of these
initiatives aim at strengthening migration control measures. This is demonstrated by
much emphasis on border surveillance projects as well as the fight against trafficking
and smuggling in North Africa. The majority of initiatives put forth considering
primary migration control strategies. As such, the EU's fortress system is designed
to illustrate just that (Fig. 2). Nonetheless, there is a need to acknowledge the EU as
part of the international community promoting peace and security through military
and civilian missions in the Sahel region. The EU fortress system does not address
the root cause of human displacement. Inopportunely, it does not highlight land
injustice and related marginalization as a migration push factor. As stated, there
appears to be an oversimplification in addressing the root cause of migration in which
most official documents highlight the need for education, creation of employment
among youths (i.e., through the promotion of entrepreneurialism), and agro-business
to address poverty [85]. A close analysis shows that different projects on managing
the root cause of irregular migration do not touch on land reform. Although there

Table 1 Mapping irregular migration management objectives as a missing agenda in the EU-Africa partnership

Key migration management objective within the agenda	EU-Africa partnership containing the agenda
Enhancing border management capacities among African countries	Khartoum Process (2004); Rabat Process (2006); Sahel Regional Action Plan 2015–2020 (2015); EU-Horn of Africa Regional Action Plan (2015)
Curbing human trafficking and migrant smuggling	Khartoum Process (2004); Rabat Process (2006); Sahel Regional Action Plan 2015–2020 (2015); EU-Horn of Africa Regional Action Plan (2015)
Strengthening of EU-Africa cooperation on return and readmission of irregular migrants	Khartoum Process (2004); Common Agenda on Mobility and Migration (2015); EU-Horn of Africa Regional Action Plan (2015); Mobility Partnership (2017)
Addressing the root cause of human displacement	Common Agenda on Mobility and Migration (2015); EU-Horn of Africa Regional Action Plan (2015); Valletta Summit Action Plan (2015)
Enhance African states' capacity to protect refugees and asylum	Common Agenda on Mobility and Migration (2015); EU-Horn of Africa Regional Action Plan (2015)

seems to lack a clear strategy from both the EU and African partners in outlining these causes, European partners expect to invest heavily in projects such as agro-business to support individual small-scale farmers [65]. It is, however, not clear how an agro-business located within a community that experiences historical land injustice and marginalization, and that suffers from chronic land conflicts, can be sustainable. Sustainable land reform ought to be the starting point in order to establish the foundation of management of second-level solutions of irregular migration (e.g., agro-business).

Despite a large amount of foreign aid to refugee-hosting countries in Africa, there are minimal environmental conservation programs that connect to the management of asylum. Broadly, this indicates challenges in designing refugee protection programs that only focus on individuals and not on habitation. All stakeholders in forced displacement and irregular migration need to put more efforts in advocacy of land policy that allows refugees and asylum seekers in Africa to engage in the production of their own food, be self-reliant, and reduce dependence on relief aid that is never enough (e.g., Uganda's recent success in achieving baseline relief in all of these categories) [86–88].

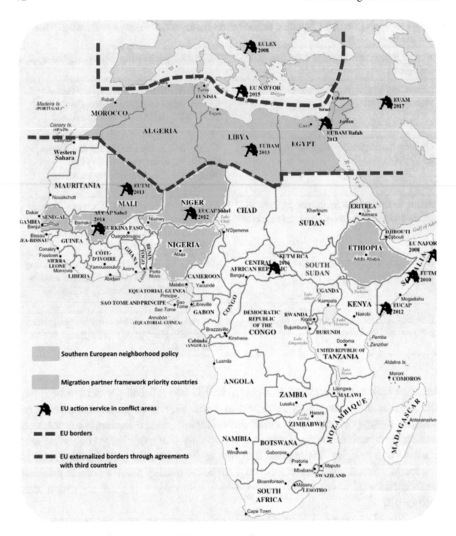

Fig. 2 EU fortress system. Adapted from the ECFR [74]

6 Concept of Sustainable Land Reform

Although there lacks a clear perspective of what the root causes of irregular migrants are, within the current international debate on addressing the root causes of irregular migration, one puzzling matter is how to achieve a sustainable strategy to address those causes [89]. Nonetheless, viewing human security as freedom from fear and freedom from want [90], land security means more than just fear of land-related conflict to include socio-ecological vulnerability. The vulnerability here includes both high exposure and low resilience to natural disasters. Lack of land (i.e., both ownership and use) or conflict over land is critical forms of vulnerability that causes irregular migration. The connection between land security and human security and stability establishes a form of sustainability. Instead of implementing conventional land reform, we need to broaden the concept to sustainable land reform under the consciousness of the current United Nations Sustainable Development Goals (SDGs) framework. This denotes efforts to establish comprehensiveness in managing irregular migration. The understanding here goes beyond the common concept of land reforms as changing the system of landownership and rights, to one that addresses the historical marginalization, sustainably. As such, the definition of sustainable land reform is a land-based development model that addresses historical injustice and socioeconomic marginalization by promoting a fairer distribution of land resources, community empowerment, and environmental security that supplements change of land use and ownership, to establish stability, resilience, and sustainability.

In this regard, sustainable land reform entails more than addressing injustice by legally defining what belongs to whom. Rather, it involves land reallocation that is accompanied by a socioeconomic stimulus package to address the effect of long-term marginalization and vulnerability as an inclusion-enhancement strategy. Therefore, sustainable land reform would free the marginalized people from economic fears such as extreme poverty and hunger as well as free communities and households from attacks by those who perceive themselves marginalized. The ability to promote land securities as an aspect of human security translates to the success in establishing a sustainable and resilient socio-ecological stability [91–93]. Vulnerability and insecurities among households are worsened by poor land planning and settling on the prohibited regions, which leave people very often exposed to natural disasters [94, 95]. On the flip side, any failure to establish sustainable land reform amounts to the maintenance of the status quo—vulnerability and human insecurity.

The sustainability of land reform calls for international development agencies to expand their scope of operation beyond a traditional reach. International development financiers and development banks in collaboration with African governments need to include sustainable land reform as one of the development areas to invest in, in an effort to establish stability and accelerate balanced economic development within African countries. The United High Commission for Refugees, International Organization for Migration, and other global organizations concerned with forced displacements and irregular migration need to take a more proactive role—supporting sustainable displacement-preventive measures including addressing the root cause of irregular migration in collaboration with sustainable land reform. By doing so, they save potential irregular migrants and asylum seekers through building resilience and addressing migration push factors sustainably.

The vast majority of African irregular migrants moving out of the continent aim to reach continental Europe. The EU and its MSs are also the most active African partner in managing irregular migration. It thus calls for an examination of how land issues (i.e., as the root causes) and sustainable land reform (i.e., as a strategy of addressing the root causes) have been articulated in the cooperation between Africa and Europe in irregular migration management. Within the EU-Africa partnership on irregular migration management, such a sustainable land reform process experiences a potential two-way challenge. From the European perspective, corruption and impunity among African elites are the major hindrances for change, especially where they are the key beneficiaries of the historical injustice. Politically connected individuals who acquired unfair chunks of land right from the colonial era are still the current holders of those lands and are still politically influential. From the African perspective, European governments lack interest to revisit the agenda of former governments' colonial actions in Africa. This leaves open the possible explanation of why sustainable land reform is a viable strategy for addressing the root causes of irregular migration but not implemented. A pessimist argument that revolves around the blame game between European governments and African elites to take responsibility is at play. Mutual understanding not to touch on the real threats (i.e., the land issue) is avoided since both would be implicated. An optimistic view is a lack of awareness of sustainable land reform as a migration management agenda between the two partners. All in all, development is not an event but a process. Sustainable land reform as a development agenda should be articulated at local, national, and international levels of the development process and in addressing the root causes of irregular migration as well as enhancing the achievement of the SDGs in Africa.

7 Conclusion

Land and land-related issues are a major cause of vulnerability, conflict, displacement, and African irregular migration to Europe. Yet, it still remains missing in the EU-Africa partnerships for irregular migration management. It is, thus, a vital area for Africa, the EU, and its MSs to explore and address the root causes of irregular migration to Europe. Setting sustainable land reform as a key agenda concern with the specific root cause-and-effect relation of irregular migration marks the beginning of addressing the real threats of potential African irregular migrants. This includes linking land reform to environmental conservation practices that replace some nonperforming strategies of environmental conservation, which are costly to maintain and do not address historical land injustices in Africa.

There is the necessity to reorient land reforms toward the establishment of socially equitable and economically viable redistribution of land in order to establish participatory and sustainable environmental conservation models that address the vulnerability and marginalization of communities in Africa. Most essentially, there is a need for inclusion of sustainable land reform as an agenda in the EU-Africa partnership on irregular migration management.

References

1. European Commission (2015) Commission Decision of 20.10.2015 on the establishment of a European Union Emergency Trust Fund for stability and addressing root causes of irregular migration and displaced persons in Africa. European Commission, Brussels
2. Crush J (2013) Between north and south: the EU-ACP migration relationship
3. Crush J (2015) The EU–ACP migration and development relationship. Migr Dev 4:39–54. https://doi.org/10.1080/21632324.2014.954377
4. Castles S, Van Hear N (2011) Root causes. In: Betts A (ed) Global migration governance. Oxford University Press, Oxford, pp 287–306
5. Docquier F, Peri G, Ruyssen I (2014) The cross-country determinants of potential and actual migration. Int Migr Rev 48:37–99. https://doi.org/10.1111/imre.12137
6. Kohnert D (2007) African migration to Europe: obscured responsibilities and common misconceptions
7. Carling J, Talleraas C (2016) Root causes and drivers of migration: implications for humanitarian efforts and development cooperation. Center for Security Studies I ETH Zurich, Peace Research Institute Oslo Publications, Oslo
8. Martin S (2010) Climate change, migration, and governance. Glob Gov 16:397–414. https://doi.org/10.2307/29764954
9. Piguet E, Pécoud A, de Guchteneire PFA (2011) Migration and climate change. UNESCO Publishing
10. Brown O (2008) Migration and climate change. Geneva
11. Castillejo C (2017) The EU migration partnership framework: time for a rethink? German Development Institute, Bonn
12. Antwi-Agyei P, Dougill AJ, Stringer LC (2015) Impacts of land tenure arrangements on the adaptive capacity of marginalized groups: the case of Ghana's Ejura Sekyedumase and Bongo districts. Land Use Policy 49:203–212. https://doi.org/10.1016/J.LANDUSEPOL.2015.08.007
13. Valipour M (2015) Land use policy and agricultural water management of the previous half of century in Africa. Appl Water Sci 5:367–395. https://doi.org/10.1007/s13201-014-0199-1
14. Boswell J, Settle C, Dugdale A (2015) Who speaks, and in what voice? The challenge of engaging 'the public' in health policy decision-making. Public Manag Rev 17:1358–1374. https://doi.org/10.1080/14719037.2014.943269
15. Frederiksen M (2016) Young men, time, and boredom in the Republic of Georgia. Temple University Press
16. Flahaux M-L, De Haas H (2016) African migration: trends, patterns, drivers. Comp Migr Stud 4:1. https://doi.org/10.1186/s40878-015-0015-6
17. Langan M (2018) Security, development, and neo-colonialism. In: Neo-colonialism and the poverty of "development" in Africa. Springer International Publishing, Cham, pp 149–175
18. Bøås M, Torheim LE (2013) The trouble in Mali—corruption, collusion, resistance. Third World Q 34:1279–1292. https://doi.org/10.1080/01436597.2013.824647
19. Andersson R (2016) Europe's failed 'fight' against irregular migration: ethnographic notes on a counterproductive industry. J Ethn Migr Stud 42:1055–1075. https://doi.org/10.1080/1369183X.2016.1139446
20. Hansen P, Jonsson S (2011) Demographic colonialism: EU–African migration management and the legacy of Eurafrica. Globalizations 8:261–276. https://doi.org/10.1080/14747731.2011.576842
21. Landau LB (2019) A chronotope of containment development: Europe's migrant crisis and Africa's reterritorialisation. Antipode 51:169–186. https://doi.org/10.1111/anti.12420
22. Mbembe A, Nuttall S (2004) Writing the world from an African metropolis. Public Cult 16:347–372. https://doi.org/10.1215/08992363-16-3-347
23. de Haas H (2014) Migration theory: quo vadis? International Migration Institute, Oxford
24. Bilgic A, Pace M (2017) The European Union and refugees. A struggle over the fate of Europe. Glob Aff 3:89–97. https://doi.org/10.1080/23340460.2017.1322252

25. Austin G (2010) African economic development and colonial legacies. Int Dev Policy 1:11–32. https://doi.org/10.4000/poldev.78
26. Rodney W (2012) How Europe underdeveloped Africa. Pambazuka Press
27. Messina J, Adhikari U, Carroll J et al (2014) Population growth, climate change and pressure on the land—eastern and southern Africa. Global Center for Food Systems Innovation, East Lansing, MI, USA
28. Lynch G (2011) The wars of who belongs where: the unstable politics of autochthony on Kenya's Mt Elgon. Ethnopolitics 10:391–410. https://doi.org/10.1080/17449057.2011.596671
29. Marshall-Fratani R (2006) The war of "who is who": autochthony, nationalism, and citizenship in the Ivoirian crisis. Afr Stud Rev 49:9–44. https://doi.org/10.1353/arw.2006.0098
30. Michalopoulos S, Papaioannou E (2011) The long-run effects of the scramble for Africa. Cambridge, MA
31. Mamdani M (2001) When does a settler become a native? Citizenship and identity in a settler society. Pretexts Lit Cult Stud 10:63–73. https://doi.org/10.1080/713692599
32. Moyo S (2018) Third world legacies: debating the African land question with Archie Mafeje. Agrar South J Polit Econ. https://doi.org/10.1177/2277976018775361
33. Moyo S (2011) Land concentration and accumulation after redistributive reform in post-settler Zimbabwe. Rev Afr Polit Econ 38:257–276. https://doi.org/10.1080/03056244.2011.582763
34. Berry SS (2018) Who owns the land? Social relations and conflict over resources in Africa. GLOCON Work Pap Ser 7:1–27
35. Anunobi F (2002) Women and development in Africa: from marginalization to gender inequality. Afr Soc Sci Rev 2:41–63
36. Ghosh D (2004) Gender and colonialism: expansion or marginalization? Hist J 47:737–755. https://doi.org/10.1017/S0018246X04003930
37. Stone R (2012) For China and Kazakhstan, no meeting of the minds on water. Science 337:405–407
38. Bayeh E (2015) The legacy of colonialism in the contemporary Africa: a cause for intrastate and interstate conflicts. Int J Innov Appl Res 3:23–29
39. Dunn KC (2009) 'Sons of the soil' and contemporary state making: autochthony, uncertainty and political violence in Africa. Third World Q 30:113–127. https://doi.org/10.1080/01436590802622417
40. Bates RH (2015) When things fell apart: state failure in late-century Africa. University of Cambridge Press, Cambridge
41. African Union (2018) Conflicts between pastoralists and farmers on the continent take more lives than terrorism. Press release no. 151/2018
42. Maitra S (2018) The changing nature of conflict: the need for a conflict-sensitive approach. In: International humanitarian action. Springer International Publishing, Cham, pp 57–78
43. Raleigh C, Linke A (2013) African conflict baselines and trends: armed conflict location and event dataset (ACLED): overview, uses & applications. Department for International Development, London, UK
44. Government of Kenya (2013) Truth, justice and reconciliation report. Government Press, Nairobi, Kenya
45. Anderson D, Lochery E (2008) Violence and exodus in Kenya's Rift Valley, 2008: predictable and preventable? J East Afr Stud 2:328–343. https://doi.org/10.1080/17531050802095536
46. Kanyinga K (2009) The legacy of the white highlands: land rights, ethnicity and the post-2007 election violence in Kenya. J Contemp Afr Stud 27:325–344. https://doi.org/10.1080/02589000903154834
47. Kanyinga K (2009) Land redistribution in Kenya. In: Agriculture land redistribution: towards greater consensus. World Bank, Washington, DC
48. Maconachie RA, Srinivasan R, Menzies N (2015) Responding to the challenge of fragility and security in West Africa: natural resources, extractive industry investment, and social conflict. World Bank, Washington, DC
49. Marc A, Verjee N, Mogaka S (2015) The challenge of stability and security in West Africa. World Bank and Agence Française de Développement, Washington, DC

50. Fon NNA (2019) Official bilingualism in Cameroon: an endangered policy? Afr Stud Q 18:55–66
51. Konings P, Nyamnjoh FB (2019) Anglophone secessionist movements in Cameroon. In: Secessionism in African politics. Springer International Publishing, Cham, pp 59–89
52. Pommerolle M-E, Heungoup HDM (2017) The "Anglophone crisis": a tale of the Cameroonian postcolony. Afr Aff (Lond) 116:526–538. https://doi.org/10.1093/afraf/adx021
53. Christopher AJ (1994) Indigenous land claims in the Anglophone world. Land Use Policy 11:31–44. https://doi.org/10.1016/0264-8377(94)90041-8
54. Stewart D (2014) What is next for Mali? The roots of conflict and challenges to stability. Strategic Studies Institute and U.S. Army War College Press, Pennsylvania
55. Langer A, Stewart F (2015) Regional imbalances, horizontal inequalities, and violent conflicts: insights from four West African countries. World Bank, Washington, DC
56. Matfess H (2017) Boko Haram: history and context. Oxford University Press
57. Olaide IA (2013) Boko Haram insurgency in Nigeria: its implication and way forwards toward avoidance of future insurgency. Int J Sci Res Publ 3:1–8
58. Turner BL, Matson PA, McCarthy JJ et al (2003) Illustrating the coupled human-environment system for vulnerability analysis: three case studies. Proc Natl Acad Sci USA 100:8080–8085. https://doi.org/10.1073/pnas.1231334100
59. UNISDR (2010) United Nations International Strategy for Disaster Reduction. United Nations, Geneva
60. Naumann G, Barbosa P, Garrote L et al (2014) Exploring drought vulnerability in Africa: an indicator based analysis to be used in early warning systems. Hydrol Earth Syst Sci 18:1591–1604. https://doi.org/10.5194/hess-18-1591-2014
61. Arnell NW, Adger WN, Thomas D, Geddes A (2013) Migration, immobility and displacement outcomes following extreme events. Environ Sci Policy 27:S32–S43. https://doi.org/10.1016/J.ENVSCI.2012.09.001
62. Hugo G (1996) Environmental concerns and international migration. Int Migr Rev 30:105. https://doi.org/10.2307/2547462
63. FAO (2017) FAO Ethiopia drought response plan and priorities in 2017, revised version. https://reliefweb.int/report/ethiopia/fao-ethiopia-drought-response-plan-and-priorities-2017-revised-version-august-2017. Accessed 3 May 2019
64. Government of Ethiopia (2018) Ethiopia: 2018 humanitarian and disaster resilience plan. GOE, Addis Ababa, Ethiopia
65. European Commission (2018) Africa-Europe alliance: first projects kicked off just three months after launch. In: EC. International Cooperation and Development. https://ec.europa.eu/europeaid/news-and-events/africa-europe-alliance-first-projects-kicked-just-three-months-after-launch_en. Accessed 3 May 2019
66. Africa-EU Partnership (2000) Africa-EU summit. In: Africa-EU summit, Cairo, 3–4 Apr 2000
67. Africa-EU Partnership (2006) Euro-African dialogue on migration and development. In: Euro-African dialogue on migration and development (Rabat process)
68. Africa-EU Partnership (2007) 2nd Africa-EU summit. In: Africa-EU partnership, Lisbon, 8–9 Dec 2007
69. Africa-EU Partnership (2010) 3rd Africa-EU summit. In: 3rd Africa-EU summit, Tripoli, 29–30 Nov 2010
70. Africa-EU Partnership (2014) 4th Africa-EU summit. In: Africa-EU partnership, Brussels, 3–4 Apr 2014
71. Africa-EU Partnership (2015) Valletta summit on migration. In: Valletta summit on migration, Valletta, Malta, 11–12 Nov 2015
72. Africa-EU Partnership (2017) 5th African Union—European Union summit. In: 5th African Union—European Union summit, Abidjan, Côte d'Ivoire, 29–30 Nov 2017
73. Africa-EU Partnership (2019) EU-AU Ministers of Foreign Affairs meeting. In: EU-AU Ministers of Foreign Affairs meeting, Brussels, 22 Jan 2019
74. ECFR (2018) The Mediterranean and migration: postcards from a "crisis". In: Migration through the Mediterranean: mapping the EU response. European Council on Foreign Relations. https://www.ecfr.eu/specials/mapping_migration. Accessed 31 July 2019

75. European Commission (2015) Implementation of the European neighbourhood policy partnership for democracy and shared prosperity with the southern Mediterranean partners report. European Commission, Brussels
76. Strik T (2017) The global approach to migration and mobility. Groningen J Int Law 5:310–328. https://doi.org/10.21827/5a6afa49dceec
77. Baldwin-Edwards M, Lutterbeck D (2019) Coping with the Libyan migration crisis. J Ethn Migr Stud 45:2241–2257. https://doi.org/10.1080/1369183X.2018.1468391
78. European Commission (2017) EUBAM-Libya initial mapping report. Working document of 24/01/2017. European External Action Service, Brussels
79. Baldwin-Edwards M, Blitz BK, Crawley H (2019) The politics of evidence-based policy in Europe's 'migration crisis'. J Ethn Migr Stud 45:2139–2155. https://doi.org/10.1080/1369183X.2018.1468307
80. Diez T, Pace M (2011) Normative power Europe and conflict transformation. In: Normative power Europe. Palgrave Macmillan UK, London, pp 210–225
81. Elgström O (2008) Images of the EU in EPA negotiations: angel, demon or just human? Eur Integr Online Pap 12:1–12
82. Elgström O (2007) Outsiders' perceptions of the European Union in international trade negotiations. J Common Mark Stud 45:949–967. https://doi.org/10.1111/j.1468-5965.2007.00755.x
83. Menéndez AJ (2016) The refugee crisis: between human tragedy and symptom of the structural crisis of European integration. Eur Law J 22:388–416. https://doi.org/10.1111/eulj.12192
84. Niemann A, Zaun N (2018) EU refugee policies and politics in times of crisis: theoretical and empirical perspectives. J Common Mark Stud 56:3–22. https://doi.org/10.1111/jcms.12650
85. European Commission (2018) Communication on a new Africa-Europe alliance for sustainable investment and jobs: taking our partnership for investment and jobs to the next level. European Commission, Brussels
86. Ilcan S, Oliver M, Connoy L (2015) Humanitarian assistance and the politics of self-reliance: Uganda's Nakivale refugee settlement. CIGI Pap Ser 86:1–18
87. World Bank (2016) An assessment of Uganda's progressive approach to refugee management. World Bank, Washington, DC
88. Smith L, Howard DA, Giordano M et al (2019) Local integration and shared resource management in protracted refugee camps: findings from a study in the Horn of Africa. J Refug Stud. https://doi.org/10.1093/jrs/fez010
89. Castillejo C (2017) The European Union Trust Fund for Africa: what implications for future EU development policy? Bonn
90. UNDP (1994) Human development report: new dimensions of human security. United Nations Development Programme, New York
91. Schilling J, Nash SL, Ide T et al (2017) Resilience and environmental security: towards joint application in peacebuilding. Glob Change Peace Secur 29:107–127. https://doi.org/10.1080/14781158.2017.1305347
92. Ide T (2017) Space, discourse and environmental peacebuilding. Third World Q 38:544–562. https://doi.org/10.1080/01436597.2016.1199261
93. Smith D, Vivekananda J (2012) Climate change, conflict, and fragility: getting the institutions right. Springer, Berlin, Heidelberg, pp 77–90
94. Mitra S, Mulligan J, Schilling J et al (2017) Developing risk or resilience? Effects of slum upgrading on the social contract and social cohesion in Kibera, Nairobi. Environ Urban 29:103–122. https://doi.org/10.1177/0956247816689218
95. Pelling M (2003) The vulnerability of cities natural disasters and social resilience. Earthscan Publications, London

Role of the International Ecological Network, Emerald, in the Western Balkans' Protected Areas

Tea Požar and Giuseppe T. Cirella

Abstract Ecological networks play an important role in controlling, preserving, and protecting nature all around Europe. Since not all European countries are part of the European Union (EU), consequently not all of the countries are part of Natura 2000, the largest European network for nature protection. All European countries, including non-EU, countries are part of one ecological network named Emerald. Emerald is international ecological network composed of Network of Areas of Special Conservation Interest (ASCI). Protected areas in Western Europe conduct more research and have more monetary assets, contrary to Eastern Europe. Research is focused on the Western Balkan countries and the number of Emerald nature protected areas on the borders of these countries. Historical overview of the political situation in this area before the Yugoslavian war is considered and consequences from this event on nature protected areas. Cross-border cooperation among many countries, cities, and villages, or even whole regions is very dependent on cooperation between geographical entities. By enlarging the Western Balkans into the EU, it is important to observe ASCI changes to the Emerald Network which presents a basis for addressing future EU-level Natura 2000 sites.

Keywords Emerald Network · Conservation · Nature protection · Western Balkans · EU

1 Introduction

The institutions of the European Union (EU) have defined the "Western Balkans" as the Balkan area that includes countries that are not members of the EU, while others

T. Požar (✉)
Institute of Geography, University of Bamberg, Bamberg, Germany
e-mail: tea.pozar@uni-bamberg.de

G. T. Cirella
Faculty of Economics, University of Gdansk, Sopot, Poland
e-mail: gt.cirella@ug.edu.pl

© Springer Nature Singapore Pte Ltd. 2020
G. T. Cirella (ed.), *Sustainable Human–Nature Relations*,
Advances in 21st Century Human Settlements,
https://doi.org/10.1007/978-981-15-3049-4_5

refer to the geographical aspects. The region includes Serbia, Bosnia and Herzegovina, Montenegro, North Macedonia, and Albania. Each of these countries' aim
is to be part of the future enlargement into the EU [1–3]. Reflecting upon the past
political situation, these countries were part of Yugoslavia, nowadays: Serbia, Bosnia
and Herzegovina, North Macedonia, and Montenegro, and an additional country to
this list is Albania, which has been an independent country since 1912. The Western Balkan countries proclaimed their independency in the following years: North
Macedonia in 1991, Bosnia and Herzegovina in 1992, and Montenegro and Serbia
in 2006 [4, 5]. According to Bennet and Wit [6], out of a total of the 38 different
ecological networking initiatives registered in Europe, three networks stand out: (1)
Pan European Ecological Network under the aegis of the Council of Europe and the
European Centre of Nature Conservation [7, 8], (2) Natura 2000 network, established
by the European Commission Habitats Directive (i.e., comprising of Special Areas
of Conservation from the Habitats Directive and Special Protection Areas from the
EU Birds Directive) [9–12], and (3) Emerald Network, also known as Network of
Areas of Special Conservation Interest (ASCI), launched in 1989 by the Council of
Europe [13]. The Emerald Network's aims are conserving important sites for nature
conservation while encouraging the protection or rehabilitation of ecological corridors between such sites. For countries that are EU members, the Emerald Network
is compatible with Natura 2000, while for non-EU countries sites are classified as
potential Natura 2000 sites. Adoption or application of the EU directives incorporates the strictest standard, in the field of nature protection [14, 15], as one of the
conditions for membership into the EU. In this chapter, the main focus will be on
the Emerald Network within the Western Balkan countries. Nominated sites from
2017 will be examined, with detailed analyses of protected areas on the borders of
the countries, to reflect the importance of transboundary cooperation.

2 The Emerald Network

The Emerald Network is an ecological network set up by the Council of Europe as
part of its work in the framework of the Bern Convention [16]. The Bern Convention
(i.e., Emerald Network) aimed to ensure the conservation of wild plants and animals
and their habitats, as an initiative of the Council [17]. It is based on recommendations
made in 1973 by the Consultative Assembly of the Council, asking for "a coherent
policy for the protection of wildlife, with a view to establishing European regulations
– if possible by means of a convention – and involving severe restrictions on hunting,
shooting, capture of animals needing protection, fishing and egg-collecting, and the
prohibition of bird netting" [16]. The final Convention not only comprises fauna,
but also flora, and came into force in 1982. The Convention falls into four parts,
including a set of appendices. Appendix I comprises a list of strictly protected flora
species, Appendix II a list of strictly protected fauna species, and Appendix III a
list of protected fauna species for which a certain exploitation is possible if the
population level permits. All species of birds (i.e., with the exception of eleven

species), amphibians, and reptiles occurring on the territories of the states that had elaborated the Convention, and not covered by Appendix II, have been included in Appendix III. The selection of species for Appendices I and II of the Bern Convention is mainly based on threat and endemism, whereas rareness is not included as a criterion [18]. Article 4 of the Bern Convention is the most relevant article, as it states that Contracting Parties "shall take appropriate and necessary legislative and administrative measures to ensure the conservation of the habitats of the wild flora and fauna species, especially those specified in Appendices I and II, and the conservation of endangered natural habitats" [19].

The creation of the Emerald Network of areas of special conservation interest was agreed in 1989 by the Standing Committee of the Bern Convention, through the adoption of Recommendation No. 16 (1989) [20] on the ASCI [21]. The Recommendation advocates Contracting Parties to take, either by legislation or otherwise, steps to designate areas of special conservation interest to ensure that necessary and appropriate conservation measures are taken for each area situated within their territory or under their responsibility [22]. All EU nature conservation legislation therefore results from the obligations of the EU as the contracting party to the Bern Convention and the consequent implementation of the 1979 European Commission Birds Directive and the 1992 European Commission Habitats Directive provided for the establishment of a representative system of legally protected areas throughout the EU, known as Natura 2000. These directives further strengthened existing protected site series at national level, or stimulated countries to define lists of protected sites.

The Standing Committee, realizing this wish and noting that the Habitats Directive was already sufficiently advanced in its work to build Natura 2000, decided to adopt its Resolution No. 3 (1996), in which it resolved to "set up a network ([i.e.,] Emerald Network) which would include the ASCI designated following its Recommendation No. 16"; it furthermore "encouraged Contracting Parties and observer states to designate ASCI and to notify them to the Secretariat" [16]. Resolution No. 3 (1996) encourages "Contracting Parties and observer states to designate ASCI" and to notify them to the Secretariat. The following forty-three European states are Contracting Parties to the Convention: Albania, Andorra, Armenia, Austria, Azerbaijan, Belgium, Bosnia and Herzegovina, Bulgaria, Croatia, Cyprus, Czech Republic, Denmark, Estonia, Finland, France, Georgia, Germany, Greece, Hungary, Iceland, Ireland, Italy, Latvia, Liechtenstein, Lithuania, Luxembourg, Malta, Moldova, Monaco, Montenegro, Netherlands, Norway, Poland, Portugal, Romania, Serbia, Slovakia, Slovenia, Spain, Sweden, Switzerland, North Macedonia, Turkey, Ukraine, UK. The following four European states have the status of observer at the meetings of the Standing Committee: Belarus, Holy See, the Russian Federation and San Marino. The participation of non-European Parties in the Emerald Network was decided by the Standing Committee in 1998. Four African states are Contracting Parties to the Convention: Burkina Faso, Morocco, Tunisia and Senegal. This rises the total to fifty-three states, which may participate in the Emerald Network [16].

The participation of states, which are not yet Contracting Parties, is not only possible, but highly desirable. Resolution No. 3 (1996) invites "European states, which are observer states in the Standing Committee of the Bern Convention, to participate in the network and designate ASCI." Resolution No. 5 (1998) establishes that for Contracting Parties that are member states of the EU, Emerald Network sites are those of Natura 2000. Indeed, no further action would be expected from them, the Natura 2000 network having identical objectives, and a more solid legal basis, to those of the Emerald Network. In this respect, the full and thorough implementation of the Habitats Directive is contemplated as a necessary and fundamental step into the achievement of the common goals it shares with the Bern Convention, both concerning the protection of natural habitats and the conservation of wild flora and fauna [16]. With the adoption in December 1998 of Resolution No. 5 (1998) "Rules for the Emerald Network," Resolution No. 6 (1998) listing the species requiring specific habitat conservation measures and the development of the bilingual version of the Emerald software, preparatory work for the launching of the Emerald Network was successfully concluded [16].

In the beginning of 1999, in order to assist the initial implementation phase of the Emerald Network, the Council of Europe proposed to a number of countries of Central and Eastern Europe to start pilot projects in their respective countries. The overall objective of the Emerald Network pilot project is to develop a pilot database, containing a fair proportion of the ASCI [21] and submit a proposal for the sites designation to the Standing Committee of the Bern Convention. In order to achieve this objective, the countries have to form project teams, carry out team training (i.e., workshops) and proceed with the scientific work (i.e., data collection on species and habitats concerned, field survey selected pilot areas, and mapping of distribution data on species and habitats), and technical tasks of installing the software and introducing site-specific data into the database (i.e., preparation of standard data sheets on the designated sites and transmitting this information in the electronic form to the Secretariat with the project report). The tasks, which are to be carried out in the framework of the Emerald Network pilot project, are described in detail in the document T-PVS/Emerald (2002) 16 "Building up the Emerald Network: A guide for Emerald Network country team leaders," which is intended as a user-friendly guide for the countries, that are implementing Emerald pilot projects [23].

An Emerald Network development program was implemented in 2005/2008, in South-Eastern Europe, as a continuation of the initial pilot projects launched by the Council of Europe. This program funded through CARDS grants [24] and thus called "the CARDS/Emerald program" targeted the following countries: Albania, Bosnia and Herzegovina, Croatia, Montenegro, North Macedonia, and Serbia. Its overall objective was to identify 100% of the potential Emerald sites, in these countries. The program benefitted from a financial contribution of the European Environmental Agency and represented an important tool contributing to preparing the countries concerned for the future work on Natura 2000 and for advance compliance with the Habitats and Birds Directives [12, 25, 26].

A joint program with the EU launched in 2009, for a period of three years, was developed to substantially develop the Emerald Network in the seven following countries: Armenia, Azerbaijan, Belarus, Georgia, Moldova, Ukraine and the European part of the Russian Federation. The objective of this program was to identify at the end of 2011 all the potential Emerald sites in the three countries of South-Caucasus and in Moldova; the objective set for Belarus and the Russian Federation amounted to 50% of the potential Emerald sites while in Ukraine, 80% of the potential Emerald sites were identified [16]. Since 8 December 2017, five countries: Belarus, Georgia, Norway, Switzerland, and Ukraine, have officially adopted Emerald sites on their territories [25]. These sites have successfully passed the biogeographical assessment for their sufficiency, as foreseen in Phase II of the Network's constitution process.

3   Western Balkan Countries: Nature Protection by IUCN and National Legislation

Categorization of the protected areas in countries of the Western Balkans will be analyzed from the scope of the International Union for Conservation of Nature's (IUCN) management of protected areas as well through each nature protection law of each country [27–29]. In accordance with the IUCN [30–32], key information regarding nature protection in the Western Balkans begins with identifying terrestrial area of protected areas in each observed country (Table 1). Based on this information, the country with highest percentage of terrestrial protected area coverage is Albania (i.e., 17.74%), while the lowest is Bosnia and Herzegovina (i.e., 1.4%). All countries included in this chapter maintain the same overall scheme of expanding their protected areas and proclaiming more nature protected areas in the future.

When considering the IUCN as the international union in the domain of nature conservation, important aspects of protection are always emphasized by IUCN's categorization of protected areas. According to the IUCN categorization, there are six classes of protected areas classified according to their management objectives. The categories are recognized by international bodies such as the United Nations and by many national governments as the global standard for defining and recording protected areas. The list of the IUCN's categorization of protected areas is as follows:

- Ia—Strict nature reserve
- Ib—Wilderness area

Table 1 Terrestrial protected area coverage of the Western Balkans

Country	Coverage (%)	Protected land area (km^2)	Total land area (km^2)
Serbia	6.61	5853	88,509
Bosnia and Herzegovina	1.4	715	51,225
Montenegro	6.4	886	13,848
North Macedonia	9.65	2456	25,443
Albania	17.74	5099	28,747

Source www.protectedplanet.net [33]

- II—National park
- III—Natural monument or feature
- IV—Habitat/species management area
- V—Protected landscape/seascape
- VI—Protected area with sustainable use of natural resources.

A breakdown of the Western Balkans' protected areas in reference to these categories is illustrated in Table 2. Utilizing this list, countries work to establish best practices and standards that maximize the effectiveness of protected and conserved areas and advance justice and equity in conservation, including the rights of indigenous peoples and local communities [34]. The categories are recognized by international bodies such as the United Nations and by many national governments as the global standard for defining and recording protected areas [35, 36] and as such are increasingly being incorporated into government legislation.

As illustrated from Table 2, the number of protected areas is highest in Serbia (i.e., 332), followed by North Macedonia (i.e., 78), while Montenegro has the least with only nine. The percentage of terrestrial protected areas coverage is led by Albania with only 61 areas listed as protected, but 17.74% of the overall protected territory. The lowest percentage of terrestrial coverage compared to land coverage is in Bosnia

Table 2 Protected areas in the Western Balkans using IUCN categories

IUCN management category	Serbia	Bosnia and Herzegovina	Montenegro	North Macedonia	Albania
Ia—Strict nature reserve	7	2	–	2	2
Ib—Wilderness area	1	2	–	–	–
II—National park	3	2	3	3	15
III—Natural monument or feature	166	11	–	57	8
IV—Habitat/species management area	34	1	–	12	22
V—Protected landscape/seascape	21	–	2	1	4
VI—Protected area with sustainable use of natural resources	2	–	–	–	4
Not reported	10	2	2	2	4
Not applicable	1	–	2	1	1
Not assigned	87	–	–	–	1
Number of protected areas	332	20	9	78	61

Source www.protectedplanet.net [33]

and Herzegovina with just 1.4%, even though still includes 20 nature protected areas. On the other hand, each Western Balkan country has its own legislative framework for nature protection designation, proclaimed from the national law of each country. According to the laws of all five Western Balkan country, it can be found that all have a clear definition of terms: protected area, ecological network, and cross-border connections of protected areas. Categorizations of protected areas differ from country to country, and similarities and differences are examined (Table 3).

- Serbia: Law on nature protection ("Official Gazette of the Republic of Serbia," No. 36/2009, 88/2010, 91/2010—correction 14/2016) [37]

Table 3 Categories of protected areas in the Western Balkans according to the nature protection laws of each country

C[a]	Bosnia and Herzegovina	Montenegro	North Macedonia	Serbia	Albania
Ia	Strict nature reserve	Strict nature reserve	Strict nature reserve	Strict nature reserve	Strict natural reserve and scientific reserve
Ib	Wildness area	–	–	–	–
II	National park	National park	National park	Special nature reserve	National park
III	Natural monument and natural features	Special nature reserve	Natural monument	National park	Natural monument
IV	Area of management of habitats and species	Nature park	Nature park	Natural monument	Managed natural reserve and natural park
V	Protected landscapes: • Landmass landscape • Sea landscape • Nature park	Natural monument	Protected landscape	Protected habitat	Protected landscape
VI	Protected areas with sustainable use of natural resources	Landscape of exceptional features	Multipurpose area	Landscape of exceptional features; nature park	Protected area of managed resources; municipal natural park; green crown

[a]IUCN management category

- Bosnia and Herzegovina: Official Gazette of the Federation of Bosnia and Herzegovina No. 66/13/28.8.2013/Law on nature protection [38]
- Montenegro: Law on nature "Official Gazette of the Republic of Montenegro," No. 054/16 from 15.08.2016 [39]
- North Macedonia: Official Gazette of the Republic of Macedonia No. 67/2004 Law on nature protection [40]
- Albania: Law No. 81/2017 for protected areas, Republic of Albania assembly [41].

Defined protected areas in the laws of Western Balkan countries include in all cases defined area and space with noticeable geological, biological, ecosystem, and landscape diversity on the territory of the defined country. All cases reflect international regulations, with the purpose of long-lasting conservation. By defining ecological network in all the laws, this definition refers to internationally connected, or spatially close, ecologically important areas which are maintained or restored to a favorable conservation status. According to all five environmental and natural laws, cross-border connection of protected areas is defined, meaning that all protected areas can connect with neighboring countries under international agreements.

From Table 3, obvious similarities can be detected, categories including strict nature reserves, national parks, natural monuments, and protected landscapes, areas, and habitats are included in all five laws of nature protection. The most similar laws are Serbian and Montenegrin which can be explained by the fact that the two countries have remained together for the longest period of time after the breakup of Yugoslavia (i.e., until 2006). Similar background laws can be explained through a complicated legal division of the protected areas in Bosnia and Herzegovina, in which the country is largely decentralized and comprises of two autonomous entities: Federation of Bosnia and Herzegovina and Republic of Srpska, with a third region, the Brčko District which is governed locally. In case of Albania, which was independent from the other four countries, the existence of additional categories, not mentioned in other laws, can be reflected, for example municipal natural park and green crown protected area with sustainable use.

4   Results

System of transboundary protected areas on borders in the Western Balkan countries will be provided from updated list of officially nominated candidate Emerald sites, from December 2017, where protected areas in following countries are proposed: Albania, Armenia, Azerbaijan, Bosnia and Herzegovina, Georgia, Republic of Moldova, Montenegro, Norway, Russian Federation, Serbia, and North Macedonia. The list of Emerald Network candidate sites is based on most recent data delivered on the Central Data Repository managed by the European Environmental Agency. It contains newly nominated sites and also previously nominated candidate sites (i.e., except those which are listed on the list of Emerald Network adopted sites). Candidate sites are defined by their code, name, and area. The list of officially nominated candidate Emerald sites is updated by the Standing Committee to the Bern Convention each year, at its annual meeting. Countries are presented in alphabetical order, and lists are sorted according to the site code in alpha-numerical order (T-PVS/PA (2017) 15) [42] (Table 4). Differences between coverage of protected areas according to the nominated candidate Emerald sites and the IUCN's list show variation. In all of the Western Balkans proposed protection is higher according to the Emerald list; in some cases, the difference is negligible, but extreme example is North Macedonia where according to the IUCN list protected area coverage is 9.65% versus the Emerald list at 29.65%. Another example is Montenegro where the percentage is more than double (i.e., 6.4% in IUCN list to 17.34% in nominated candidate Emerald list).

In Table 5, nominated candidate Emerald sites on the borders are analyzed. The highest number of protected areas along borders can be found in Serbia and Montenegro (i.e., 28, the same number of sites). The lowest number of nominated candidate Emerald sites is Bosnia (i.e., 10). When considering percentage of coverage of nominated candidate Emerald sites, country with the highest percentage of coverage is North Macedonia with 22.45%, and one with the lowest 2.67% is Bosnia and Herzegovina.

Table 4 Nominated candidate Emerald sites in the Western Balkans

Country	Number of protected areas	Coverage (%)	Land area protected (km^2)	Total land area (km^2)
Serbia	61	11.54	10,210.78	88,509
Bosnia and Herzegovina	29	4.74	2427.32	51,225
Montenegro	32	17.34	2400.77	13,848
North Macedonia	35	29.65	7543.83	25,443
Albania	25	18.17	5224.29	28,747

Table 5 Nominated candidate Emerald sites in border areas in the Western Balkans

Country	Number of protected areas	Coverage (%)	Land area protected (km^2)	Total land area (km^2)
Serbia	28	6.55	5793.83	88,509
Bosnia and Herzegovina	10	2.67	1375.92	51,225
Montenegro	28	16.19	2242.59	13,848
North Macedonia	24	22.45	5713.15	25,443
Albania	19	15.63	4494.13	28,747

Source T-PVS/PA (2017) 15 [42]

Illustrative maps for the nominated candidate Emerald sites for each country of the Western Balkans are listed. Figures use the same criteria of buffer zones (i.e., 20 km from the international borderline). In the case of Bosnia and Herzegovina, Montenegro and Albania, Emerald sites which are positioned along coastal buffer zones are also included as border protected areas. Listed nominated candidate Emerald sites are as follows:

- Serbia (Fig. 1): Gornje Podunavlje; Prokletije; Vlasina; Sar Planina; Tara; Stara planina; Djerdap; Ludasko jezero; Zasavica; Dolina Pcinje; Suboticka pescara; Vrsacke planine; Sargan-Mokra Gora; Pasnjaci velike droplje; Selevenjske pustare; Klisura reke Milesevke; Palic; Zlatibor; Jerma; Pester; Karadjordjevo; Klisura reke Tresnjice; Venerina padina; Tesne jaruge; Zelenicje; Zelenika; Tikvara; and Zaovine

- Bosnia and Herzegovina (Fig. 2): Kompleks Maglic-Volujak-Zelengora; Popovo polje and Vjetrenica; Pecine kod Brckog; Raca-Bijeljina; Bardaca-Lijevce polje; Fatnicko polje; Gatacko Veliko polje; Veliki Stolac; Kanjon Drine; and Livanjsko Polje

- Montenegro (Fig. 3): Maglic, Volujak, and Bioc; Canyon of Mala Rijeka; Durmitor mountain with Tara River Canyon; Skadar Lake; Velika Plaza with Solana Ulcinj; Buljarica; Field Cemovsko polje; Kanjon Cijevne; Lovcen; Tivatska solila; Sasko jezero, rijeka Bojana, Knete, and Ada Bojana; Rumija; Cave in Djalovica Ravine; Plavsko-Gusinjske Prokletije (+ Bogicevica); Lim River; Valley of Cehotina River; Ljubisnja; Golija and Ledenice; Ostatak kanjona Pive ispod Hidroelektrane; Visitor and Zeletin; Kotarsko Risanski Bay; Orjen; Pecin beach; Hajla; Spas, Budva; Komovi; Katici, Donkova, and Velja Seka Islands; and Platamuni

- North Macedonia (Fig. 4): Galichica; Ezerani; Dojransko Ezero; Pelister; Mavrovo; Shar Planina; Matka; Smolarski Vodopad; Monospitovsko blato; Alshar; Nidze; Kozuf; Jablanica; Belasica; Blato Negorski banji; Ohridsko Ezero; Prespansko Ezero; Osogovski Planini; Churchulum (Bogdanci); German-Pchinja; Klisura na Bregalnica; Mariovo; Maleshevski Planini; and Gorna Pelagonija

- Listed nominated candidate Emerald sites in border area of Albania (Fig. 5): "Llogara" National Park; Divjaka National Park; Prespa National Park; Butrinti National Park; Allamani; Protected landscape of the wetland complex Vjose-Narte; Managed Nature Reserve (i.e., the Albanian part) of Shkodra Lake; Alps; Morava; Karaburun-Orikum-Dukat National Park; Karavasta National Park; Shengjin-Ishem; Pogradec Protected Landscape; Managed Nature Reserve Germenj-Shelegure-Leskovik-Piskal; Protected Landscape of Buna River-Velipoja; National Park Rrajce-Shebenik; Protected Landscape of Korabi; Managed Nature Reserve Rrushkulli-Ishem; and Managed Nature Reserve of Berzane.

Based on the results from the individual figures of the Western Balkan countries, one unique figure with all five Western Balkan countries is provided (Fig. 6) where additional results can be discussed. Based on spatial distribution of nominated candidate the Emerald sites in border areas of the Western Balkans with the highest level of density of protected sites (i.e., at least six), four focus regions are emphasized:

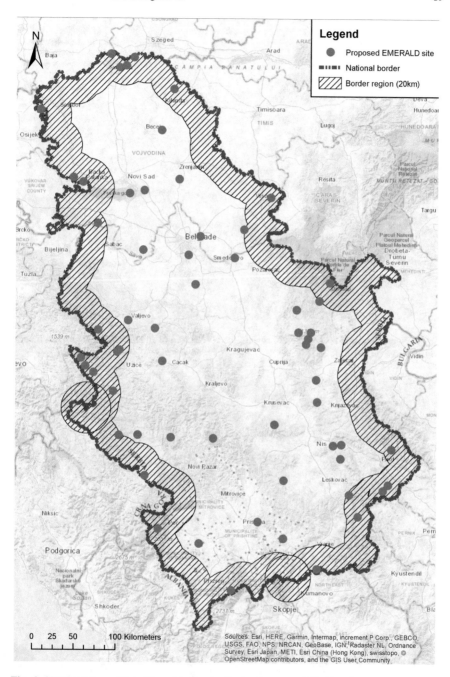

Fig. 1 Nominated candidate Emerald sites in the border area of Serbia. *Source* Authors own elaboration, 2018

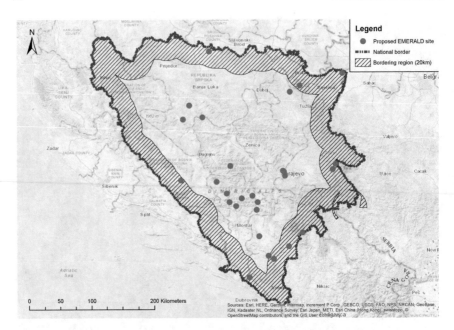

Fig. 2 Nominated candidate Emerald sites in the border area of Bosnia and Herzegovina. *Source* Authors own elaboration, 2018

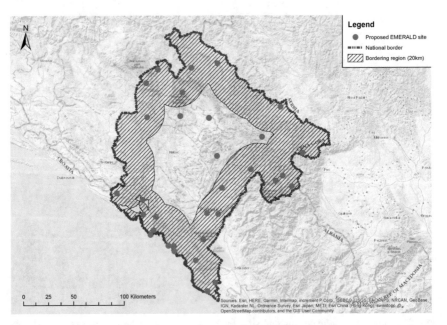

Fig. 3 Nominated candidate Emerald sites in the border area of Montenegro. *Source* Authors own elaboration, 2018

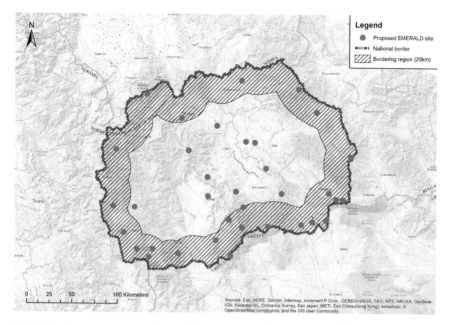

Fig. 4 Nominated candidate Emerald sites in the border area of North Macedonia. *Source* Authors own elaboration, 2018

- border of Serbia with Bosnia and Herzegovina (i.e., Tara, Sargan-Mokra Gora, Zlatibor, Klisura reke Tresnjice, Tesne jaruge, Zelenika and Zaovine on Serbian side, and Veliki Stolac and Kanjon Drine on side of Bosnia and Herzegovina)
- tri-border area of Serbia, Montenegro, and Albania (i.e., Prokletije in Serbia, Alps in Albania and Hajla, Lim River, Visitor, Zeletin and Plavsko-Gusinjske Prokletije (+ Bogicevica) in Montenegro)
- border of Montenegro with Albania (i.e., Skadar Lake, Rumija, Sasko jezero, rijeka Bojana, Knete, Ada Bojana and Velika Plaza with Solana Ulcinj in Montenegro and Managed Nature Reserve (i.e., the Albanian part) of Shkodra Lake and Protected Landscape of Buna River-Velipoja in Albania)
- border of Albania with North Macedonia (i.e., Galichica, Ohridsko Ezero, Prespansko Ezero and Pelister in North Macedonia and Prespa National Park and Pogradec Protected Landscape in Albania).

Within the focus regions, direct cross-border protected sites can be detected: Dolina Pcinje-German-Pchinja (i.e., Serbia-North Macedonia); Sar Planina-Shar Planina (i.e., Serbia-North Macedonia); Maglic, Volujak, and Bioc, Ostatak kanjona Pive ispod Hidroelektrane-Kompleks Maglic-Volujak-Zelengora (i.e., Montenegro-Bosnia and Herzegovina); Cave in Djalovica Ravine-Pester (i.e., Montenegro-Serbia); Protected Landscape of Korabi-Mavrovo (i.e., Albania-North Macedonia); and National Park Rrajce-Shebenik-Jablanica (i.e., Albania-North Macedonia).

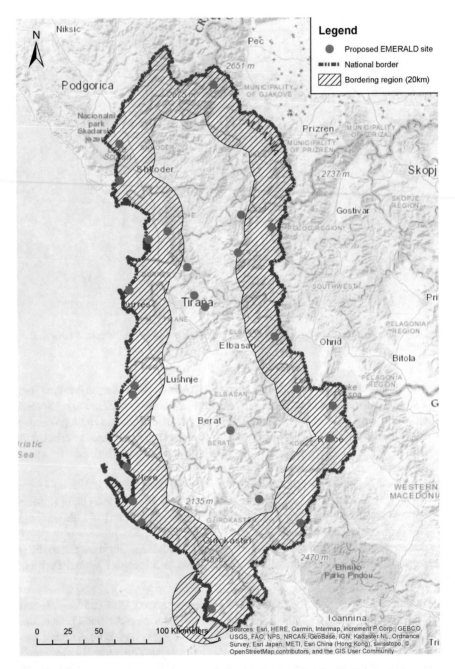

Fig. 5 Nominated candidate Emerald sites in the border area of Albania. *Source* Authors own elaboration, 2018

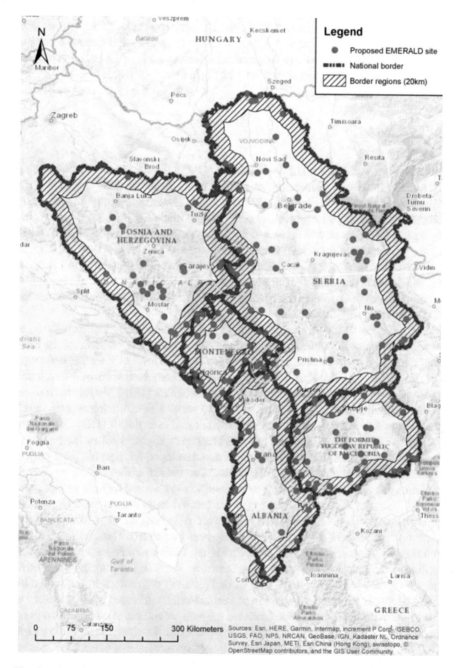

Fig. 6 Nominated candidate Emerald sites in the border area of the Western Balkan countries with focus regions. *Source* Authors own elaboration, 2018

The largest nominated candidate Emerald site is Prokletije in Serbia with 1553.96 km^2 of protected area, together with the Albanian Alps and Montenegrin Plavsko-Gusinjske Prokletije (+ Bogicevica) represent a unique and sole example of a tri-border protected area in the region with the total protected area of 2486.12 km^2.

5 Discussion

Increasing the vulnerability of the living world, with the constant trend of reducing the number of species, Europe has responded with a determined resolve to stop the reduction of biodiversity by establishing a single legal framework and ecological network as a basis for the protection policy nature in the EU [8, 14, 43]. According to European Environmental Agency data, eleven biogeographical regions are spread across the European continent, where four of them are on territory of the Western Balkan countries: Alpine, Continental, Mediterranean, and Pannonian [44]. Biogeographical regions are used to distinguish broad geographical areas that have a similar set of climatic, topographic, and geological conditions. This in turn influences the kind of animals and plants that can be found [45].

The situation in the Western Balkans can be observed from earlier periods, in which the Federal Republic of Yugoslavia location, centers upon some of the most important biogeographical regions in Europe. The high level of scientific education and university expertise has, for many years, been combined with well-developed awareness—and concern—to protect natural resources [46]. This further has led to the fact that Montenegro declared itself the world's first "environmental state," pledged to develop a harmonious relationship with nature at the 1992 Rio Conference [47]. In terms of nature protection, after the breakup of Yugoslavia, first interventions could be observed in Serbia, where they have developed an example pilot project "Emerald Network in the Republic of Serbia." Coordinator of the project, the Faculty of Biology at the University of Belgrade, identified species and habitats proposing to expand the list of Emerald species important for Serbia, as well as the selection and description of potential Emerald sites nationwide. Six areas (9.83%) of the total sixty-one proposed potential "Emerald" areas have been completed. These include Gornje Podunavlje, Kopaonik, Obedska bara, Prokletije, Deliblatska Pescara, and Vlasina. In the second phase of the project, implemented in 2006, the remaining fifty-five areas (90.16%) were completed. The second phase of the project, implemented by the Institute for Nature Conservation of Serbia [48], coordinated by the ministry at that time in charge of these issues, revised the number and status of certain areas on the List of potential Emerald sites in Serbia which were developed in the 2005 phase. The third phase of the project, among other things, revised the lists of priority habitats, identified species present in Serbia, according to EU directives, and provided data on their populations at the national level.

According to available data in other Western Balkan countries, for example, North Macedonia, agreements of cooperation in the field of environmental protection with Albania and Greece have been signed. On February 2, 2001, World Wetlands Day,

Prime Ministers from all three countries declared "Prespa Park" a new trilateral trans-boundary protected area. At this signing, it was recognized that conservation of nature depends largely on respect for international legal agreements that aim at protecting natural environments [49]. Detailed description about the eventual project cannot be found for all countries of the Western Balkans. In Montenegro, however, a project establishing the foundations for Natura 2000 network was submitted for funding to the European Commission through the Instrument for Pre-accession Assistance 2012–2013. Also, there is an on-going project "Integrated management of Skadar Lake ecosystem" between Montenegro and Albania on the largest lake throughout the Balkan Peninsula.

Transboundary issues are of critical importance to biodiversity conservation in many countries, means that greater regional cooperation needs to be encouraged. The importance of transboundary issues can be dated back to 1998 when Brunner [50] pointed out that four out of Serbia's five National parks (i.e., including the proposed Prokletije National Park which abuts Albania) are situated on boundary areas. Kopaonik National Park is also situated partly in Kosovo. Two of Montenegro's four parks are also on the frontier with Albania. In some areas, particularly those such as Lake Skadar which abuts with Albania has little or no cooperative agreement with adjoining conservation authorities. By contrast collaboration with Hungary in respect of the Selevenj sands which adjoins to Kiskunsag National Park is cited as a model for transboundary collaboration with a joint management plan and proposals to facilitate cross-border crossings. From these examples, we can see that transboundary collaboration is possible throughout the territory of the Western Balkans. Examples that are described as negative are between countries that are core of this chapter, while positive examples mentioned cooperation between Serbia and neighboring Hungary.

6   Conclusion

European initiative and ecological networks should be used as examples of trans-boundary cooperation for a European Green belt initiative. This initiative serves to harmonize the system of nature protected areas in Europe, which spread across 23 countries and connect some of the most different biodiversity systems. It is espe-cially important since it deals with protected areas in border regions emphasizing cooperation regionwide.

The Emerald Network is currently in its implementation phase [17, 21, 25]—one of the most critical steps. It is important to indicate the development of the Emerald Network-General Viewer, which visually shows the overview of Emerald Network, both adopted and candidate sites [51]. When considering countries of the West-ern Balkans, territories of Serbia, Bosnia and Herzegovina and North Macedonia are included in the Viewer. Emerald Network-General Viewer is established by the Council of Europe and European Environmental Agency. For the Western Balkan countries, it is important to underline that in many cases problems in project real-ization are defined in strategic documents. The absence of a national environmental strategy, as one reason for the gap of clear priorities for the country's international environmental cooperation, is common. Institutional capacity of the environmental authorities is weak with insufficient level of project documentation. In most cases, there is no adjustment to many of the segments of legislation within the EU Directives, as such a lack of financial assets and human resources needs to be properly implemented for future sustainability-oriented conservation.

References

1. Schimmelfennig F, Sedelmeier U (2019) The Europeanization of Eastern Europe: the external incentives model revisited. J Eur Public Policy 1–20. https://doi.org/10.1080/13501763.2019. 1617333
2. Richter S, Wunsch N (2019) Money, power, glory: the linkages between EU conditionality and state capture in the Western Balkans. J Eur Public Policy 1–22. https://doi.org/10.1080/ 13501763.2019.1578815
3. Anastasakis O (2008) The EU's political conditionality in the Western Balkans: towards a more pragmatic approach. SE Eur Black Sea Stud 8:365–377. https://doi.org/10.1080/ 14683850802556384
4. Jano D (2008) From "Balkanization" to "Europeanization": the stages of Western Balkans complex transformations. L'Europe Form 349–350:55. https://doi.org/10.3917/eufor.349.0055
5. Ker-Lindsay J, Armakolas I, Balfour R, Stratulat C (2017) The national politics of EU enlargement in the Western Balkans. SE Eur Black Sea Stud 17:511–522. https://doi.org/10.1080/ 14683857.2017.1424398
6. Bennett G, Wit P (2001) The development and application of ecological networks: a review of proposals, plans and programmes. AID Environ
7. Jongman RHG, Bouwma IM, Griffioen A et al (2011) The Pan European ecological network: PEEN. Landsc Ecol 26:311–326. https://doi.org/10.1007/s10980-010-9567-x
8. Jones-Walters L (2007) Pan-European ecological networks. J Nat Conserv 15:262–264. https:// doi.org/10.1016/J.JNC.2007.10.001
9. Popescu VD, Rozylowicz L, Niculae IM et al (2014) Species, habitats, society: an evaluation of research supporting EU's Natura 2000 network. PLoS ONE 9:e113648. https://doi.org/10. 1371/journal.pone.0113648
10. Gruber B, Evans D, Henle K et al (2012) "Mind the gap!" How well does Natura 2000 cover species of European interest? Nat Conserv 3:45–63. https://doi.org/10.3897/ natureconservation.3.3732
11. Hochkirch A, Schmitt T, Beninde J et al (2013) Europe needs a new vision for a Natura 2020 network. Conserv Lett 6
12. Paavola J (2004) Protected areas governance and justice: theory and the European Union's habitats directive. Environ Sci 1:59–77. https://doi.org/10.1076/evms.1.1.59.23763
13. COE (2018) Emerald Network of Areas of Special Conservation Interest
14. Friedrichs M, Hermoso V, Bremerich V, Langhans SD (2018) Evaluation of habitat protection under the European Natura 2000 conservation network—the example for Germany. PLoS ONE 13:e0208264. https://doi.org/10.1371/journal.pone.0208264
15. Díaz S, Fargione J, Chapin FS, Tilman D (2006) Biodiversity loss threatens human well-being. PLoS Biol 4:e277. https://doi.org/10.1371/journal.pbio.0040277
16. Council of Europe (2009) Convention on the conservation of European wildlife and natural habitats, Standing Committee (2009). The Emerald Network of Areas of Special Conservation Interest. Fact sheet 20
17. Bevz O (2018) Legal regulation of the Emerald Network: national and global aspects. J Vasyl Stefanyk Precarpathian Natl Univ 5. https://doi.org/10.15330/jpnu.5.2.91-98

18. Ozinga WA, de Heer M, Hennekens SM et al (2005) Target species—species of European concern: a database driven selection of plant and animal species for the implementation of the Pan European ecological network. Alterra, Wageningen

19. Council of Europe (2010) Convention on the conservation of European wildlife and natural habitats, Standing Committee: criteria for assessing the national lists of proposed Areas of Special Conservation Interest. Council of Europe, Brussels

20. Council of Europe (1989) Convention on the conservation of European wildlife and natural habitats, Standing Committee (1989) recommendation no. 16 (1989) of the Standing Committee on Areas of Special Conservation Interest. Appendix 5 to the Emerald Network

21. Council of Europe (2003) Group of experts for the setting up of the Emerald Network of Areas of Special Conservation Interest: report T-PVS/Emerald (2003) 11. Council of Europe, Brussels

22. Council of Europe (2010) Convention on the conservation of European wildlife and natural habitats, Standing Committee (2010). Biogeographical regions' map. T-PVS/PA (2010) 14

23. Council of Europe (2002) Convention on the conservation of European wildlife and natural habitats, Standing Committee (2002). Building up the Emerald Network: a guide for Emerald Network country team leaders. T-PVS/Emerald (2002) 16

24. Council of Europe (2007) Development of the Emerald sites network in the West-Balkan under the CARDS program: report. Council of Europe, Brussels

25. Council of Europe (2016) Group of experts on protected areas and ecological networks: report T-PVS/PA(2016) 4. Council of Europe, Brussels

26. Pajević A (2006) Emerald Network pilot project in Serbia and Montenegro: project report T-PVS/Emerald01e_06. Council of Europe and Montenegrin Ministry of Environmental Protection and Physical Planning, Podgorica

27. IUCN (2018) Protected areas. https://www.iucn.org/theme/protected-areas. Accessed 2 Feb 2019

28. IUCN (2008) International Union for Conservation of Nature website. http://www.iucn.org/. Accessed 13 Dec 2017

29. IUCN (2018) Invasive alien species. IUCN, Brussels

30. Rodríguez JP, Keith DA, Rodríguez-Clark KM et al (2015) A practical guide to the application of the IUCN Red List of Ecosystems criteria. Philos Trans R Soc Lond B Biol Sci 370:20140003. https://doi.org/10.1098/rstb.2014.0003

31. Rodríguez JP, Rodríguez-Clark KM, Keith DA et al (2012) IUCN Red List of Ecosystems. SAPIENS 5:61–70

32. Keith DA, Rodríguez JP, Rodríguez-Clark KM et al (2013) Scientific foundations for an IUCN Red List of Ecosystems. PLoS ONE 8:e62111. https://doi.org/10.1371/journal.pone.0062111

33. Protected Planet (2018) Protected area coverage per country and territory by UN environment regions

34. Dudley N, Shadie P, Stolton S (2013) Guidelines for applying protected area management categories including IUCN WCPA best practice guidance on recognising protected areas and assigning management categories and governance types. IUCN, Gland, Switzerland

35. Boitani L, Cowling RM, Dublin HT et al (2008) Change the IUCN protected area categories to reflect biodiversity outcomes. PLoS Biol 6:e66. https://doi.org/10.1371/journal.pbio.0060066

36. Lopoukhine N, Crawhall N, Dudley N et al (2012) Protected areas: providing natural solutions to 21st century challenges. SAPIENS 5:117–131

37. Official Gazette of the Republic of Serbia (2018) Law on nature protection (No. 36/2009, 88/2010, 91/2010—correction 14/2016)

38. Official Gazette of the Federation of Bosnia and Herzegovina (2018) No. 66/13/28.8.2013/ Law on nature protection

39. Official Gazette of the Republic of Montenegro (2018) Law on nature, No. 054/16 from 15.08.2016

40. Official Gazette of the Republic of North Macedonia (2018) No. 67/2004 Law on nature protection

41. Republic of Albania (2018) Law No. 81/2017 for protected areas

42. Council of Europe (2017) Convention on the conservation of European wildlife and natural habitats, Standing Committee (2017). Updated list of officially nominated candidate Emerald sites. T-PVS/PA (2017) 15
43. Nesshöver C, Vandewalle M, Wittmer H et al (2016) The network of knowledge approach: improving the science and society dialogue on biodiversity and ecosystem services in Europe. Biodivers Conserv 25:1215–1233. https://doi.org/10.1007/s10531-016-1127-5
44. European Topic Centre on Biological Diversity (2006) The indicative map of European biogeographical regions: methodology and development. European Environment Agency Press, Paris
45. Rueda M, Rodríguez MÁ, Hawkins BA (2010) Towards a biogeographic regionalization of the European biota. J Biogeogr 37:2067–2076. https://doi.org/10.1111/j.1365-2699.2010.02388.x
46. Blaženčić J, Stevanović V, Vasić V (1995) Biodiverzitet Jugoslavije sa pregledom vrsta od međunarodnog značaja. Ecolibri, Beograd
47. Carter FW, Turnock D (1998) Yugoslavia in environmental problems in east central Europe. Routledge, London
48. ZZPS (2018) Institute for nature conservation of Serbia. Emerald. http://www.zzps.rs/novo/index.php?jezik=en&strana=zastita_prirode_ekoloske_mreze_emerald. Accessed 1 Feb 2019
49. USAid (2001) EandE—117/119 report—Macedonia I: introduction. USAid, Washington, DC
50. Brunner R (1998) Transborder protected area cooperation in Yugoslavia: a technical assessment. Vienna: report on a visit to Serbia. In: Carter FW, Turnock D (eds) Yugoslavia in environmental problems in east central Europe. Routledge, London, pp 396–416
51. Emerald Network (2018) General Viewer. http://emerald.eea.europa.eu/. Accessed 5 Feb 2019

Economic Evaluation: Perspective Ideas

How Efficient is Urban Land Speculation?

Bedane Sh. Gemeda, Birhanu G. Abebe, and Giuseppe T. Cirella

Abstract Urban land speculation presents planning challenges of contemporary urbanization and settlements worldwide. Studies concerning land speculation focus primarily on the Global North and Asia, while little has been done on Sub-Saharan Africa. Available research in Sub-Saharan Africa is largely confined to studying economic forces driving peri-urbanization, land markets, and informal influences. Few have explicitly examined the policy forces driving it. Urban areas in Ethiopia have been growing very quickly in recent decades, which have led to ever-increasing demand for land in peri-urban and urban areas for housing and other non-agricultural activities. This has had several transformative impacts on transitional peri-urban areas, including engulfment of local communities and conversion of land rights and use from an agricultural to a built-up property rights system. Peri-urban areas also compete for land among speculators of diverse backgrounds. This chapter analyzes the urbanization and policy forces driving land speculation, economic role of land speculators, opportunity of land speculation and motives behind land speculation in the city of Shashemene, Ethiopia. It scrutinizes the city's urbanization policy and national land policy by investigating how and why they are linked with the city's land speculation processes. The analyses utilize primary data collected through household surveys, field observations, and key informant interviews, which are complemented by secondary data from national legal and policy documents, and regional and city administrative reports. The chapter encompasses an attempt to discover the process of land speculation in peri-urban and urban land by looking at the principal actors involved in land speculation. A principle finding illustrates that the societal costs in

B. Sh. Gemeda (✉) · B. G. Abebe
Ethiopian Institute of Architecture, Building Construction and City Development, Addis Ababa University, Addis Ababa, Ethiopia
e-mail: bedanes@yahoo.com

B. G. Abebe
e-mail: mesi.bire@gmail.com

G. T. Cirella
Faculty of Economics, University of Gdansk, Sopot, Poland
e-mail: gt.cirella@ug.edu.pl

© Springer Nature Singapore Pte Ltd. 2020
G. T. Cirella (ed.), *Sustainable Human–Nature Relations*,
Advances in 21st Century Human Settlements,
https://doi.org/10.1007/978-981-15-3049-4_6

a typical city are increased by about 5–11% as a result of speculative increases in the value of urban land.

Keywords Speculators · Urbanization · Tax model · Opportunity cost · Ethiopia

1 Introduction

The incidence of land speculation is seen as far-reaching [1, 2]. The border area of Latin America has been categorized by land speculation and development [3–5]. In the same way, in the Chinese Pearl River Delta, economic growth induces a boom in property values which motivated the speculation of land [6–8]. According to Baird [9], there is a need to identify the experience associated with and punishment of land speculation. In the USA, unused housing (i.e., speculated housing) creates offshoots of urban slums and motivates the growth of suburbanization [10–12]. During China's period of land reform, elevated land prices triggered land speculation and became a source of revenue [7]. In Hong Kong, land speculation has led to a dynamic property market. The increasing tendency in property sales revealed Hong Kong's economy as entwined with a range of sectors (e.g., banking, construction, real estate stocks, and transportation). This economic effect has had a long-lasting effect [13]. In Bangkok, evidence confirms that surplus purchasing of housing by land speculators resulted in over-investment by developers matching unsustainable and unrealistic demand levels, thereby disrupting markets. This turbulence force 350,000 housing units to be built with only one-third of them purchased by residents [14, 15].

The habitual economic description of speculative market activities concerns the motives of buying and selling. An absentee land owner is one who deals with a certain good in order to make use of expected change in its price [16–19]. The good is neither consumed or used in production while in the possession of the speculator. In 1939, Kaldor [20] provided the definition speculation can "be defined as the acquire (or trade) of goods with a vision to re-trade (or re-acquire) later on, where the reason at the back of such action is the prospect of a change in the pertinent prices relative to the ruling price and not a gain accruing through their use, or any kind of alteration effected in them or their transfer between dissimilar markets." This explanation works fine for money, goods, and other markets in which "holding" and "using" are jointly exclusive activities. However, if holding a good for future sale at a gain does not prevent using it in the interim, then this definition is inadequate. As such, it is somewhat inappropriate for speculation in durable goods (e.g., real property) which can make available a stream of useful services over time while being held for a capital gain [1, 21]. If we decide to define speculation with a definition that precludes simultaneous uses, as in Kaldor's definition, we effectively disallow durable goods speculation and narrow the field of activities which are actually speculative. This chapter investigates three key basics of this problem, they include: (1) determine the effect of population growth on developing land value, (2) derive speculative reaction to the growth in land value, and (3) relate social costs for the city and speculative bidding for uptown land using a resource distribution model.

2 Literature Review

Land speculation has been intertwined with land expansion throughout much of the development of the USA. Historians have criticized the widespread occurrence of speculative land sales in the development of the American border throughout the nineteenth century [22], even though speculative land sales had been generally prejudice based the two centuries prior. There have been a number of scenario-driven examples of speculative fever throughout much of the world's colonial era. Some argue that land speculation was a chief stirring force behind the development of the West; however, notwithstanding the fact that land speculation has become an American ritual, it can also be said that it has often been negatively maligned [2, 23]. For example, Yearwood [24] stated how speculation can lead to the misuse of valuable resources. He explains: "there is waste in taking land out of productive use before it is ready for another; squander in tying up capital for long periods in a barren enterprise, with more in taxes, interest, and extraordinary assessments; worst in the division of land into lots that are too small, or of poor design, or poorly to be found; waste in zoning too much for business use; and waste in replotting land which has been precipitated subdivided. There are other wastes too, increased overall government costs, all of which affect the community unfavorably."

It is doubtful that land speculation is as spiteful as some believe; there are some well-documented studies that examine its effects. Three items relating to speculation in urban land appear to be well understood: (1) uptown land pending conversion for urban use increases in price, (2) such land will eventually be sold to developers for many times the value as opposed to agricultural use, and (3) large areas of the land will remain inactive at any given point in time [25]. As such, Schmid [26] presented data which revealed that in the period 1958–1964, the price of land for development in the USA increased five times faster than the consumer price and three times the construction cost. Clonts [4] predicted the land price of Virginia, at the outer edge of Washington, D.C., and found that variables related to urban development explained 30% of the deviation in the value of land and improvements. Adams et al. [27] studied the movement of "underdone" land prices in the Philadelphia edge between 1948 and 1962; they found that after adjusting for various characteristics of the land, prices rose between 13 and 16%, respectively. Their findings stated "long-run tendency of prices is close to a normal return in accord with the notion that the development trend of Northwest Philadelphia was anticipated early and capitalized into real estate values." That is, the verification is consistent with what would be expected in a well-functioning market [28].

In a historical sense, the association between the value of uptown versus agricultural land has been looked at by Maisel [29] and Schmid [26]. Maisel reported that land value in San Francisco was ten times greater than that of agricultural land, whereas Schmid reported values of land were ten times greater than agricultural values in a number of other selected American cities. As such, we find that speculative activity increased land values by a factor of ten in comparison with the agricultural value. However, in any given city, the rate of population growth, income, and other factors would affect the relationship between uptown land and agricultural value. Expanding upon this notion, Clawson [10] estimated that about half the land in American cities was idle at any given point. Given the verification that suburban land values have risen rapidly while much land is idle, some important questions remain. Is speculative activity in charge for what is observed? How does this rapid rise in price affect resource allocation? The large amount of idle land in cities, often described as urban sprawl or leapfrogging, is universally ascribed to speculation. As such, it is plausible that speculation carried out by rational profit maximizers, in a well-functioning market, could support for much of the idle land.

Eliminating obfuscating factors, by assuming land to be harmonized in all respects except distance to employment, it is doubtful that profit maximization would result in adjacent plots being sold at disparate times to developers. Rather, leapfrogging is more likely to be due to imperfections in the market. First, the land market during a thin market period renders sales tricky and its accuracy unstable. For example, a farmer or a speculator who receives an attractive offer may sell ahead of time if they believe another offer may not be forthcoming. Second, there is substantial uncertainty among speculators expected future income (e.g., factors include changes in governmental policy and population growth). As a result, profit-maximizing speculators can easily differ in their expectations and sell at different times. Third, the most advantageous holding period is influenced by the discount rate applied to the income stream. Speculators may face differing costs of funds which can lead to different holding periods. Fourth, high taxes can drive inadequately capitalized speculators (e.g., farmers) out of the market impulsively; thus, even though costs of a slump and leapfrogging may be substantial, speculation, per se, may not be the cause. Actions to improve the functioning of the market rather than the elimination of speculators may be sufficient to reduce sprawl [10, 30].

In terms of the effect of high land prices on resource distribution, Neutze [31] argues the notion of misallocation. "There appears to be no reason, in terms of resource allotment, why the cheapest land that is regarded as being urban should exceed its opportunity cost. Other land will be more expensive because of the differential value accorded to land that is favorably located for development as a result. If land prices are too high in this sense, development will occur at too high density in order to economize on land, too much amounts of resources will be used in building." In a severely static context Neutze is correct; however, in a growing city where redevelopment is sporadic, initially it may be desirable to build higher density in order to postpone redevelopment due to coupled costs to a later date. Uptown land prices would give confidence to higher density development. As a supplement, several authors argued that certain resources are underpriced in urban areas [23, 32, 33]; henceforth, most attention has been drawn on public services, pollution, and congestion of the transport system. If in actuality, these resources are underpriced, too numerous people will be attracted to cities. On the other hand, if built-up land is priced above its opportunity cost, lands, and therefore housing, are overpriced—people will be repelled from cities. This overpricing might be a second best that offsets underpricing elsewhere; thus it is possible that by bidding up the price of land speculators are improving their resource share.

3 Methodology and Research Approach

Research methodology is a way of systemizing observations, describing methods of collecting evidence, and indicating the types of tools to be used during data collection [34]. There are several types of research strategies (e.g., case studies, surveys, experiments, ethnography, phenomenology, grounded theory, action research, and archival analysis) that land speculation research focuses on [35, 36]. Strategies provide an alternative way of collecting and analyzing empirical evidence. It is informed by three main factors, namely type of research questions posed, extent of control over the actual behavioral events, and degree of focus on contemporary events [37]. Based on the quantitative and qualitative nature of the research, a combination of three approaches is used: (1) case study, (2) desk review, and (3) model application. The existing theoretical and analytical frameworks such as the property speculation analytical framework and institutional analyses were also employed to understand, examine the nature of, and develop theories and concepts about urban land speculation.

3.1 Case Study Approach

An empirical inquiry investigates complex and contemporary societal phenomena with the intention of focalizing on a specific sample study [34]. The approach can predominantly be employed in relation to the discovery of information following inductive logic (i.e., to describe what is happening in the case study setting) [35, 37–39]. The case study approach is also helpful when "how" and "why" questions are being posed and when the investigator has little control over events in a situation too complex for surveys or experimental research [40, 41]; thus, it creates an opportunity to combine different data collection techniques such as interviews, observations, questionnaires, focus group discussions (FGDs), and document analyses. Thus, Shashemene, the capital city of West Arsi Zone, was selected as the case-study area based on a purposive non random sampling technique. The reason why Shashemene was selected is that it is one of the fastest-growing urban centers in Oromia both demographically and socially.

3.2   Desk Review Research Approach

Desk review is a research approach that primarily entails a literature review. Using the desk research approach, we explored the existing issues with facts obtained from a wide variety of secondary data sources [40]. The main purpose of using this method was to develop a conceptual framework to examine the nature of land speculation in peri-urban areas. The method was also used to capture best practices of land development and administration that could be applicable to the peri-urban contexts of the study area. Hence, accordingly, books, articles, policy papers, reports and records of Shashemene City Land Agency were reviewed to consolidate the findings.

3.3   Methods

3.3.1   Case Study Area

Shashemene is a city and a separate district in West Arsi Zone, Oromia Region, Ethiopia. It is about 250 km away from the capital city of Addis Ababa. It has a latitude of 7° 12′ north and a longitude of 38° 36′ east. Shashemene is surrounded by lands in all directions and covers a total surface area of about 12,994.61 ha. The distance is about 12 km from its north to south extreme outer edges and 8 km from east to west [42] (Fig. 1). In order to understand and analyze the wider context of land speculation in an era of rapid urbanization, Ethiopia-wide, Shashemene, and its surrounding peri-urban *kebeles* (i.e., villages) were selected since its urban center is one of the fastest growing in Ethiopia, both demographically and spatially. The current population, including peri-urban areas, is estimated to be about 120,000; however, this number is expected to grow fourfold by 2040 [43]. The built-up area within the city's jurisdiction is also expected to expand at an even faster rate than its population.

According to the Shashemene 2005 Master Plan, seven peri-urban rural and agricultural *kebeles* are located immediately outside its municipal boundary [44]. Among the peri-urban *kebeles*, per-urban Awasho, per-urban Bulchana, and the urban *kebele* of Abosto were selected as case study areas.

3.3.2   Model Specification

This research adopts an easy model of resource sharing proposed by Capozza et al. [45] for urban areas. The model's assumptions include: (1) the city is rectangular with width and length; (2) at one end of the city is the service center, and the rest is completely residential; (3) employees go back and forward to work from the central business district (CBD) on the transportation scheme which requires no surface

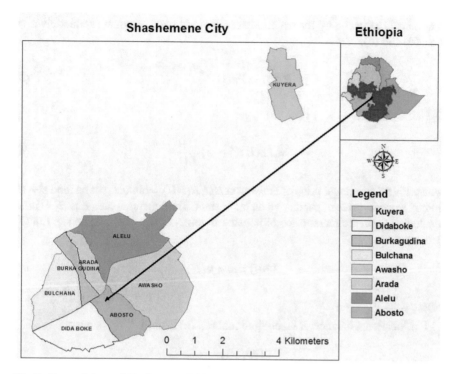

Fig. 1 Size and shape of Shashemene, Ethiopia

land; (4) transport cost is straight with remoteness and equivalent to 's' in Ethiopian Birr (i.e., where 1 Ethiopian Birr = US$ 0.035) per kilometer; and (5) distance is normalized with the border of the CBD equaling zero with one employee per family and each family requiring one unit of accommodation. The extension of the city is due to the augmentation of inhabitants. The ratio of workers to the total population is constant, as in Eq. (1).

$$N = N(t) \tag{1}$$

where N = number of employees and 't' = time. Equation (1) defines the number of personnel at each point in time. Accommodation is formed by using capital and land. Changeover of factors is possible at any time (i.e., the model is self-motivated in which development can take place gradually but steadily without loss of efficiency). Accordingly, this is developed by utilizing Eq. (2).

$$H(U) = \overline{A} L(U)^a K(U)^{1-a} \tag{2}$$

where $H(U)$ = accommodation output U, kilometers from the CBD, $L(U)$ = land used in housing at U, $K(U)$ = capital used in housing at U, and $\overline{A}$ and 'a' are constants. The necessity for which factors are paid their marginal product gives us Eqs. (3) and (4).

$$R(U) = Ph(U)^a \frac{H(U)}{L(U)}, \tag{3}$$

and

$$r = Ph(U)(1-a)\frac{H(U)}{L(U)}, \tag{4}$$

where $Ph(HU)$ = rental price of housing at HU, $R(HU)$ = annual rent on land at HU, and 'r' = rental price of capital which is constant. The equations (i.e., Eqs. 2–4) infer a calculation on the rental price of housing in terms of factor prices; hence, Eq. (5) is needed.

$$Ph(U) = AR(U)^a \tag{5}$$

where $A = \frac{r^{1-a}}{\overline{A}a^a(1-a)^{1-a}}$.

For locational balance, it is required that Eq. (6) is used.

$$\frac{d[Ph(U)]}{du} = -S \tag{6}$$

Adding Eqs. (5) and (6) and solving for land rentals gives us:

$$R(U) = \left(\frac{\cdot c - su}{A}\right)1/a,$$

where 'c' is a constant of the amalgamation. If
$\widehat{R} = R(\hat{U}) = \left(\frac{c-s\hat{U}}{A}\right)1/a$, Eq. (7) can be calculated as such:

$$R(U) = \left[\widehat{R}^a + \frac{s}{A}(\hat{U} - U)\right]1/a \tag{7}$$

where U = remoteness to the edge of the city at 't' = time. As such, all workers must be accommodated in the city; hence, Eq. (8) is used.

$$\int_0^{\hat{U}} H(U)du = N(t) \tag{8}$$

Solving Eqs. (3) and (5) for H, using the actual city perimeter, the unit width is calculated so that $L(U) = 1$, as well as substituting the result in Eq. (8), Eq. (9) is formulated.

$$\int_0^{\hat{U}} R \frac{R(U)^{1-a}}{Aa} du = N(t) \tag{9}$$

Further formulation can be applied by substituting from Eq. (7) by integrating and rearranging the notation; Eq. (10) is calculated as the radius of the city as a function of the population.

$$\hat{U}(t) = \frac{A}{S}\left\{\left[\widehat{R}+sN(t)\right]^a - \widehat{R}^a\right\} \tag{10}$$

Equations (7) and (10) then can be used to settle the land rent structure of the city over space and time, formulating Eqs. (11) or (12).

$$R(V,t) = \left\{[\widehat{R}+sN(t)]^a - \frac{sU}{A}\right\}1/a \tag{11}$$

or

$$\ln R = \frac{1}{a}\ln\left\{[\widehat{R}+sN(t)]^a - \frac{sU}{A}\right\} \tag{12}$$

3.3.3 Data Collection Methods and Analysis

One of the principal advantages of using the case study approach is the ability to use mixed and multiple sources of data. As such, both quantitative and qualitative data were collected from primary and secondary data sources, including key informant interviews (i.e., both structured and open-ended), direct field observations, questionnaires, FGDs, and document analyses. The research employed a mixture of quantitative and qualitative data analysis techniques so as to capture the complex and multifaceted reality of land speculation. Quantitative data collected through questionnaires were analyzed through simple descriptive statistics using percentages and means. The quantitative technique is aimed to quantify land speculation using model specification. Qualitative data was analyzed using triangulation, concepts and opinion interpretation, and comparing and contrasting methods. Such data were presented in the form of text.

4 Results and Discussion

4.1 Developed Value of Land

The alteration of land for urban use is a highly complex process involving a large number of private individuals as well as private and public institutions (i.e., further detailed can be found in Clawson [10]). As such, the following scenario can be abstracted from the complexities. Four principle roles are played by a variety of actors, including homebuyer, developer, land investor (i.e., speculator), and farmer (i.e., rural landowner). In the first stage, land is held by the rural landowner. Land use is stable and land rent is a constant (i.e., R_o). In the second stage, the city's expansion begins to consume nearby vacant agricultural land. Investors then see the possibility of higher prices for land for urban use and begin bidding up land values. Conversion takes place after the land is sold to developers. The developer holds the land only briefly while housing is under construction. The process is concluded when the land and improvements are sold to homebuyers.

For any given plot of land, one actor may play more than one role. Whether or not plots of land actually change hands is unimportant. The essential feature is the ability and willingness of investors to accept a negative cash flow for a period of time in anticipation of future capital gains. As such, at any given point, Eq. (7) describes the rental value of land over space; this relationship is shown in Fig. 2. One question that remains is what homebuyers will be willing to pay for the developed land. If homebuyers are lucid participants with perfect forethought, they should be ready to pay a price for the land equal to the present value of future rent. This of course would be a severe supposition which requires clients be aware of differential appreciation

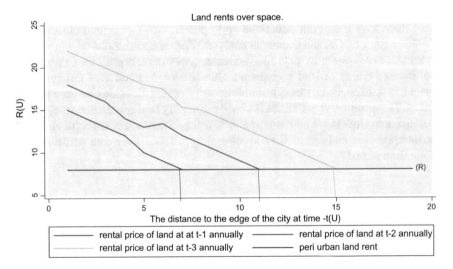

Fig. 2 Land rent over space

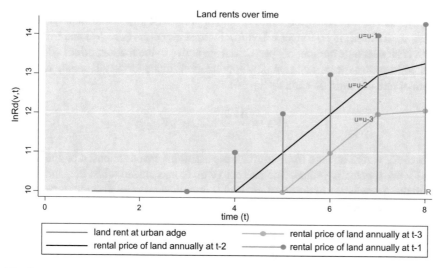

Fig. 3 Land rent over time

rates between locations. Homebuyers do appear to be conscious of the appreciation potential of land, but casual empiricism indicates that there is a little consciousness, if any, among appreciation rates variability in a systematic manner with regards to location.

A more realistic postulation about the operation of the housing market would be that buyers contrast the current rents, taxes, and values among locations as well as have a willingness to pay the capitalized value of the current rent and less the current taxes. In other words, the value of developed land to homebuyers utilizes Eq. (13), where 'v' is the tax rate.

$$V\mathrm{d}(v, t) = \frac{R(v, t)}{r + v} \tag{13}$$

The developed value of land will follow a trajectory similar to that in Fig. 3, where the function is convex for small exponential growth rates (i.e., convex growth). If the city is growing linearly, the logarithm of land rent and value at any given location will increase at a decreasing rate. In this case, a finite speculative price is guaranteed as long as the ultimate growth rate is less than the discount rate.

4.2 Speculative Value of Land

Given the rate of appreciation of developed land (i.e., $V\mathrm{d}(V, t)$), speculators with perfect prescience will offer the price of land until appreciation is equivalent to holding costs. That is, if the land value is rising swiftly in price, speculators can craft capital gains from buying and holding until appreciation plus rent from undeveloped (e.g., agriculture) is no longer offsetting interest expenses (i.e., 'r') and taxes (i.e., 'v'). If an adequate number of speculators are active in the market, prices will rise to the point where the current rent plus appreciation minus taxes will equate to only a normal rate of return on capital (Eq. 14).

$$rVs = \frac{dVs}{dt} + R_o - vVs \tag{14}$$

where R_o = rent on land for non-urban use assuming taxes are based on the market and Vs = speculative value. The present value is maximized when the land is held until the appreciation of the developed value is equivalent to interest plus taxes minus undeveloped rentals as formulated in Eq. (15).

$$\frac{dVd}{dt} = (r + v)Vd - R_o \tag{15}$$

The development date is denoted td, as shown in Fig. 4, which explains the condition where R_o is insignificant. Speculators, as a result, will begin to offer the price of land at 'u' above its opportunity cost (i.e., Vo), long before the city reaches that plot of land. Adding up the land will be withheld from development for a period of time (i.e., $t1 - to$), longer than in the nonattendance of speculation. Land is first developed at price versus $Vo = \left[\frac{R_o}{(r+V)} \right]$.

Notice in Fig. 5 that an increase in the tax rate from $v1$ to $v2$ would seem to decrease the speculative value of land and move the date of development forward. However,

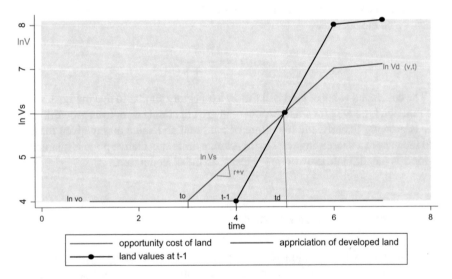

Fig. 4 Speculative price

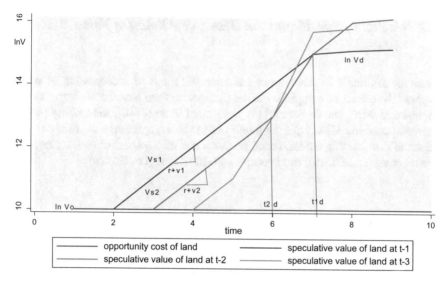

Fig. 5 Effect of tax changes on the speculative price

this neglects the impact of the tax rate on developed values. In fact, if buyers value land by discounting the excess of rental value over an allowance for taxes according to $Vd = \frac{1}{r}(R - vVd)$, or $Vd = \frac{R}{r+v}$, then Eq. (15) is to be formulated as Eqs. (16) or (17).

$$R - R_o = \frac{1}{r+v}\frac{dR}{dt} \tag{16}$$

or

$$r + v = \frac{dR}{dt}\frac{1}{R - R_o} \tag{17}$$

As such, if the right side is decreasing over time, increasing 'v' advances the date at which this condition is met. Another effect must also consider if tax is applied for improvements in which case $r + v$ must replace 'r' in Eq. (17), foremost to lessen capital intensive development, abiding to the former outward movement of the edge of the city and earlier development of any given land sale.

In summary, speculators lift the price of land to developers above the value of the land during the status of undeveloped use and put off the date of development; this, in effect, causes smaller pieces of land a higher rate of tax. This result diverges from the usual proposition that land tax value is neutral; as noted, it solely depends on the assumption of the buyer that the tax will remain fixed at the level of the date of purchase, whereas the seller allows for increases in the tax if he holds onto the land for a longer period.

4.3 Upshot of Speculation on Resource Allotment Value of Land

Resource allotment in urban areas has been the focus of a succession of recent papers. The effects of congestion in the transport system have been a key concern, for instance, Mills and de Ferranti [3], Solow and Vickrey [46], and Livesey [47]. A universal constituent has been the inquiry of the best city through the minimization of social cost. In this model, social costs consist of transport costs (i.e., Eq. 18), capital costs (i.e., Eq. 19), and opportunity cost of land (i.e., Eq. 20).

$$\int_0^{\hat{U}} sUH(U)du \tag{18}$$

$$\int_0^{\hat{U}} rK(u)du \tag{19}$$

$$\int_0^{\hat{U}} R_o u\,du = R_o \hat{U} \tag{20}$$

where R_o = rent on land in an option, non-urban use. Social costs at a given point in time are then given by the sum of equations Eqs. (18), (19), and (20), namely (21).

$$C(t) = \int_0^{\hat{U}} suH(u) + rK(u) + R_o u\,du \tag{21}$$

where $C(t)$ = total social cost at 't' = time. Integrating notation from the speculative value of land Eq. (22) is formulated.

$$C(t) = \int_0^{\hat{U}} \left[\left(\frac{su}{aA} \right) Ru^{1-a} + \frac{1-a}{a} R(u) \right] du + R_o \hat{U} \tag{22}$$

As such, we know that land will be supplied to developers at the speculative price, so Eq. (23) can be derived.

$$\widehat{R} = Rs(V, td) \tag{23}$$

In addition, it is possible to vary the price of land at the edge of the city (i.e., R) by varying the tax rate (i.e., 'v'), as defined in Eq. (24).

$$\Delta = \widehat{R} - R_0 \qquad (24)$$

where Δ = best market over the opportunity cost of land which must be paid to speculators. This best market notation takes on values as the tax rate varies; thus, the best-paid speculators can decrease or be ultimately eliminated by an appropriate tax policy. An appropriate tax policy is the one which minimizes the present value of social costs (i.e., $\Delta(t)$), such that the notation from Eq. (25) is minimized, where 'p' is the social rate of discount.

$$\int_0^T C(t) \exp(-pt) dt \qquad (25)$$

This is a degenerate problem in the calculus of variations, the solution of which is independent of time. A closed-form solution has not been obtained, but numerical experimentation yields the result that a minimum exists using Eq. (26) in which $\Delta = 0$.

$$\widehat{R} = R_0 \qquad (26)$$

Thus, social costs are minimized if the whole speculative value above opportunity costs are taxed away.

4.4 Arithmetical Result: Social Cost of Speculation

A valid critic of the model is that some issues concerning social cost of speculation cannot be answered; however, additional aftereffects can be obtained by using arithmetical methods. Of particular, interest is the question of how robust an increase in social costs is caused by speculation. To this end, a numerical analysis of a model similar to that presented can be applied. The key distinction is that in the simulated model, the city is a two-dimensional circular city rather than the less realistic linear city considered earlier—if not the assumptions would be indistinguishable. The transport cost is linear in distance, and housing is produced by using capital and land in accordance with Eq. (22). The experimental city was taken to have 2046 workers in the CBD, with the majority of other parameters used by Capozza et al. [45]. One omission, however, was the factor share of land in the production of housing. Total social costs are responsive to this parameter, requiring a precise value in accordance with the previous research. A number of studies suggest the likely range is from 15 to 25% [33]. To bracket the probable values, simulations were done with the factor share of land at each end of the range. The greater the factor share of land, the larger the loss of speculative bidding for land. One final value is important—the relationship between the opportunity cost of land (i.e., R_0) and the speculative supply price of land to developers (i.e., $\widehat{R}$). Maisel [29] estimates the price of land to developers to be about ten times the price of agricultural land, agriculture being taken to be the best alternative use. Accordingly, within the study area, the land to developer result is found to be about 10.3 times the price of agricultural land—very similar with Maisel's study. Social costs are tabulated for this range of speculative values and for the two values of the factor share of land, illustrated in Table 1. Note that speculation, which increases land values ten times its opportunity cost, will increase social costs by 5.6% (i.e., with a factor share of land at 15%) and 11.3% (i.e., with a factor share of land at 25%).

Table 1 Social costs for values of $\widehat{R}$ with a factor share of land at 15 and 25%, values in parentheses are percentage differences from values when $\widehat{R} = \widehat{R}_0$

Factor share of land	Speculative value of land $(\widehat{R})$	Radius of the city	Opportunity cost of land[a]	Capital costs[a]	Transport costs[a]	Total social costs[a]
15%	0	77.7 (715)	17.3 (44.1)	162.4 (−2.5)	56.1 (1.3)	235.8 (0.8)
	R_0	45.3	12.0	166.6	55.4	234.0
	2 R_0	42.0 (−7.6)	11.6 (−3.33)	169.2 (1.7)	43.9 (−20.9)	234.7 (0.3)
	4 R_0	38.4 (−0.15)	11.3 (−5.8)	173.3 (4.3)	51.8 (−6.5)	236.3 (1.2)
	6 R_0	36.3 (−25.5)	11.1 (−7.5)	176.6 (6.4)	50.2 (−11.9)	237.8 (1.9)
	8 R_0	34.7 (−23.4)	11.0 (−8.3)	179.1 (7.5)	48.8 (−14.5)	237.8 (1.9)
	10 R_0	33.6 (−25.8)	10.9 (−9.2)	187.5 (12.5)	48.7 (−12.0)	247.1 (5.6)
25%	0	159.6 (58.1)	45.8 (97.4)	269.6 (−8.1)	156.5 (5.3)	471.9 (6.02)
	R_0	100.9	23.2	293.4	148.6	445.1
	2 R_0	91.7 (−9.2)	20.7 (−10.8)	307.0 (4.6)	141.4 (−4.8)	469.0 (5.4)
	4 R_0	81.9 (18.8)	18.3 (−21.1)	326.4 (11.3)	131.4 (11.6)	476.0 (6.9)
	6 R_0	76.1 (−27.6)	17.0 (−26.7)	341.6 (16.4)	124.5 (−16.2)	483.0 (8.5)
	8 R_0	71.9 (−28.7)	16.1 (−30.6)	354.0 (20.7)	119.2 (−19.2)	489. 5 (9.9)
	10 R_0	68.8 (−31.9)	15.5 (−33.2)	365.3 (24.6)	114.9 (−22.7)	495.7 (11.3)

[a]Represents thousands of Ethiopian Birr per day (i.e., 1 Ethiopian Birr = US$ 0.035)

As the speculative value of land rises, the radius of the city decreases due to increased population density. The opportunity cost of land also decreases as less land is used. Capital costs, on the other hand, rise due to diminishing returns since more capital is applied to the same land. Transport costs decrease as R increases, since workers commute shorter distances in the more compact cities. Total social costs decrease then rise, and a turning point is reached when land is valued at its opportunity cost.

One study indicates that active speculation based on public investments may have predated these initiatives. According to a number of respondents in the study, sales of vacant land surrounding the proposed industrial park observed that annual sales tripled in the three years following the initial industrial park announcement (i.e., 2005–2007). The three-year land assembly phase can be recognized as a potential for increased social and economic costs, resulting in market distortion and land speculation.

Long-term speculative strategies are also common throughout the study area. Interviewees confirmed that it is customary for vacant lot owners to hold land for decades waiting until land prices rise or to return to the previous levels after a market downturn. Other long-term speculators noted local individuals and families looking at vacant land as a financial asset, similar to long-term bonds. Reports from the quantitative data confirm that many vacant properties have been singularly owned for decades, which increase value of developed and speculative land, in some cases placed in family trusts. Not only are trusts especially reluctant to sell land for immediate development, but also sometimes properties are buried in family holdings, becoming a low priority among a variety of financial assets. Finally, some long-term speculators are local individuals highly engaged in a speculation-based business model where properties may be rented for billboards or as temporary construction storage space while land prices appreciate.

4.5 Improving Resources

The improvement of resources is a causation of land speculation. By bidding up the price of land, land speculators can allocate resources among citizens. This can be annotated with the following six assumptions: (1) households do not do as they are told in terms of low of demand and no consumer substitution, (2) in spite of the price of housing, each household occupies a standard dwelling, with 160 m^2 of living space, (3) typical households have a fixed monthly cost of 1513 Ethiopian Birr (i.e., US$ 53) required to spent on rent and commuting each month, (4) cost of commuting is 120 Ethiopian Birr (i.e., US$ 4.25) per month per 500 m from CBD, (5) price of housing is defined as the price per square foot (i.e., 0.093 m^2) of housing per month, and (6) as such, if a household rents a 178-square-foot house (i.e., a standard house) for 1000 Ethiopian Birr (i.e., US$ 34.50), the price of housing is 5.6 Ethiopian Birr (i.e., US$ 0.20) per square foot (i.e., calculated by dividing 1000 Ethiopian Birr by 178 square feet).

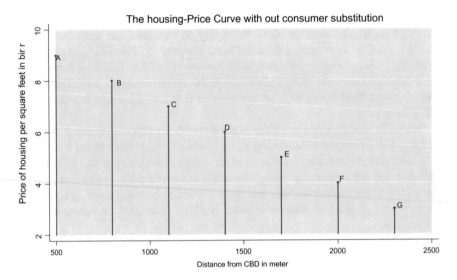

Fig. 6 Housing-price curve without consumer substitution based on the case study

Expanding upon this scenario, a symmetrical illustration is put together to illustrate the housing-price curvature (Fig. 6). For the typical residence, situated next to the service area, commuting cost is zero and families would be able to utilize the full 1513 Ethiopian Birr (i.e., US$ 53) on accommodation, paying 8.5 Ethiopian Birr (i.e., US$ 0.30) per square foot for a 178-square-foot dwelling (Fig. 6, Point A). As such, commuting cost is equal to 120 Ethiopian Birr (i.e., US$ 4.15) per 300 m from the service area. To best understand the logic of a negatively sloped housing-price curve, deem what would come about if it were horizontal, with a constant price of 5.6 Ethiopian Birr (i.e., US$ 0.20) per square foot at all locations in the city. For a family, living 2000 m from the employment area, a move to a location next to the service area would eliminate 480 Ethiopian Birr (i.e., US$ 16.50) of commuting cost without any modification in housing costs. Other households would have the same incentive to move closer to the service area; hence, the demand for housing near the employment center would increase, pulling housing and land prices. Concurrently, the demand would decrease at more remote locations, pushing down prices. In other words, a horizontal housing-price curve would be transformed into a negatively sloped one.

The symmetrical housing-price curve makes inhabitants indifferent to location. A move away from the employment area changes commuting cost by the change in distance (i.e., Δx) multiplied by the commuting cost per meter (i.e., 't') and changes housing cost by the change in price of housing (i.e., Δp) multiplied by housing consumption (i.e., 'h'). For locational indifference, two changes must sum to zero as in Eq. (27).

$$\Delta p \cdot h + \Delta x \cdot t = 0 \tag{27}$$

We can rewrite this expression as Eq. (28) to show that the change in housing cost equals the negative of the change in commuting cost.

$$\Delta p \cdot h = -\Delta x \cdot t \tag{28}$$

We can also use a trade-off expression to get an equation for the slope of the housing-price curve by dividing each side of the expression by Δx and 'h' as in Eq. (29). Specifically, decreasing housing costs must offset increased transport cost as distance to the CBD increases. Implicit in the assumption is that each household demands have an affixed quantity of housing.

$$\frac{\Delta p}{\Delta x} = \frac{t}{h} \tag{29}$$

Thus, the overpricing of land may improve resource allocation by offsetting some of the underpricing elsewhere, and as a result, cities may be closer to optimizing size.

5 Conclusion

We have demonstrated that speculation in urban land raises the price of land for developers and postpones the date of the development. This speculative ordering for land influences resource distribution by causing land to be developed at a date after the best market scenario. The option that an urban resource (i.e., in this case "land") is highly priced due to speculation is mostly exciting. A great deal of the literature in urban economics argues the opposite, namely that due to overcrowding, effluence, or other diseconomies, resources are underpriced. This, the overpricing of land, may advance resource distribution by offsetting some of the underpricing elsewhere; hence, as a consequence, cities may develop to the best market size earlier. It would be practical for future research to deem how the pricing of one resource affects another and how social costs are influenced by a variety of relating circumstances.

It has been revealed that land predicament, in the study area, is land speculation. Land speculation has resulted in market malfunction, both in efficiency and equity. Repeated land speculation increases have been based on three myths about urban land: (1) the legend of land shortage, (2) the myth of continuous increase in land price, and (3) the legend of the futility of land policies. The basic reasons for rapid increases in land prices are the swiftness with which the demand for urban uses has grown, the confines placed by the government on rural-to-urban alteration, and the private appreciation of socially created land value. It has been argued, moreover, that a practicable and moderately simple land policy does exist and can be used to chase away these three myths. To dampen the speculative demand for land and to apt communally the increases in value of land, the effective tax rate of the global land value tax should be raised, and all loopholes in the real estate capital gains tax closed. As a result, sustainable economic valuation of the land-use control system should be rehabilitated to endorse rural-to-urban conversion while still promoting investment-friendly development.

References

1. Dye RF, England RW (2010) Assessing the theory and practice of land value taxation. Lincoln Institute of Land Policy, Cambridge, MA, USA
2. Hoyt H (2000) One hundred years of land values in Chicago: the relationship of the growth of Chicago to the rise of its land values, 1830–1933. Beard Books
3. Mills ES, de Ferranti DM (1971) Market choices and optimum city size. Am Econ Rev 61:340–345. https://doi.org/10.2307/1911814
4. Clonts HA Jr (1970) Influence of urbanization on land values at the urban periphery. Land Econ 46:489. https://doi.org/10.2307/3145522
5. Smith J, Cadavid J, Rincón A, Vera R (1997) Land speculation and intensification at the frontier: a seeming paradox in the Colombian savanna. Agric Syst 54:501–520. https://doi.org/10.1016/S0308-521X(96)00088-1
6. Yeh A, Xia L (1999) Economic development and agricultural land loss in the Pearl River Delta, China. Habitat Int 23:373–390. https://doi.org/10.1016/S0197-3975(99)00013-2
7. Du J, Peiser RB (2014) Land supply, pricing and local governments' land hoarding in China. Reg Sci Urban Econ 48:180–189. https://doi.org/10.1016/J.REGSCIURBECO.2014.07.002
8. Wang H, Wu X, Wu D, Nie X (2019) Will land development time restriction reduce land price? The perspective of American call options. Land Use Policy 83:75–83
9. Baird IG (2014) The global land grab meta-narrative, Asian money laundering and elite capture: reconsidering the Cambodian context. Geopolitics 19:431–453. https://doi.org/10.1080/14650045.2013.811645
10. Clawson M (1962) Urban sprawl and speculation in suburban land. Land Econ 38:99–111. https://doi.org/10.4324/9781315823911-38
11. Lindeman B (1976) Anatomy of land speculation. J Am Inst Plann 42:142–152. https://doi.org/10.1080/01944367608977715
12. Stanley BW (2016) Leveraging public land development initiatives for private gain. Urban Aff Rev 52:559–590. https://doi.org/10.1177/1078087415579733
13. Chui E (2001) Doomed elderly people in a booming city: urban redevelopment and housing problems of elderly people in Hong Kong. Hous Theory Soc 18:158–166. https://doi.org/10.1080/14036090152770528
14. Pornchokchai S, Perera R (2005) Housing speculation in Bangkok: lessons for emerging economies. Habitat Int 29:439–452. https://doi.org/10.1016/J.HABITATINT.2004.01.002
15. Gul A, Nawaz M, Basheer MA et al (2018) Built houses as a tool to control residential land speculation—a case study of Bahria Town, Lahore. Habitat Int 71:81–87. https://doi.org/10.1016/J.HABITATINT.2017.11.007
16. Bayer PJ, Geissler C, Roberts JW, Roberts JW (2011) Speculators and middlemen: the role of flippers in the housing market. Working paper 16784
17. Joshua PB, Glanda GG, Ilesanmi FA (2016) The effects of land speculation on urban planning and development in Bajabure Area, Girei Local Government, Adamawa State. J Environ Earth Sci 6:128–133
18. Wubneh M (2018) Policies and praxis of land acquisition, use, and development in Ethiopia. Land Use Policy 73:170–183. https://doi.org/10.1016/J.LANDUSEPOL.2018.01.017
19. Thontteh EO, Babarinde JA (2018) Analysis of land speculation in the urban fringe of Lagos, Nigeria. Pac Rim Prop Res J 24:161–184. https://doi.org/10.1080/14445921.2018.1461770
20. Kaldor N (1939) Speculation and economic stability. Rev Econ Stud 7:1. https://doi.org/10.2307/2967593

21. Malpezzi S, Wachter SM (2005) The role of speculation in real estate cycles. J Real Estate Lit 13:143–164. https://doi.org/10.2307/44103516
22. Loehr RC (1971) Pioneers and profits: land speculation on the Iowa frontier. Narnia
23. Goodfellow T (2015) Taxing the urban boom: property taxation and land leasing in Kigali and Addis Ababa. SSRN Electron J. https://doi.org/10.2139/ssrn.2634036
24. Yearwood RM (1968) Land: American attitudes on speculation, development and control. Ann Public Coop Econ 39:215–224. https://doi.org/10.1111/j.1467-8292.1968.tb00209.x
25. Archer RW (1973) Land speculation and scattered development: failures in the urban-fringe land market. Urban Stud 10:367–372
26. Schmid AA (1968) Converting land from rural to urban uses. Resources for the Future
27. Adams FG, Milgram G, Green EW, Mansfield C (1968) Undeveloped land prices during urbanization: a micro-empirical study over time. Rev Econ Stat 50:248. https://doi.org/10.2307/1926200
28. Lin R, Zhu D (2014) A spatial and temporal analysis on land incremental values coupled with land rights in China. Habitat Int 44:168–176. https://doi.org/10.1016/J.HABITATINT.2014.06.003
29. Maisel SJ (2005) Price movements of building sites in the US: a comparison among metropolitan areas. Pap Reg Sci 12:47–60. https://doi.org/10.1111/j.1435-5597.1964.tb01252.x
30. Carruthers JI, Ulfarsson GF (2003) Urban sprawl and the cost of public services. Environ Plan B Plan Des 30:503–522. https://doi.org/10.1068/b12847
31. Neutze M (1970) The price of land for urban development. Econ Rec 46:313–328. https://doi.org/10.1111/j.1475-4932.1970.tb02492.x
32. Thomas R, Reed M, Clifton K et al (2018) A framework for scaling sustainable land management options. Land Degrad Dev 29:3272–3284. https://doi.org/10.1002/ldr.3080
33. Mills ES (1968) Studies in the structure of the urban economy. Resources for the Future, Washington
34. Cavaye ALM (1996) Case study research: a multi-faceted research approach for IS. Inf Syst J 6:227–242. https://doi.org/10.1111/j.1365-2575.1996.tb00015.x
35. Denscombe M (2010) The good research guide: for small-scale research projects. Open University Press, Berkshire, United Kingdom
36. Kitzinger J (1995) Qualitative research: introducing focus groups. BMJ 311:299–302. https://doi.org/10.1136/bmj.311.7000.299
37. Yin RK (2003) Case study research: design and methods. Sage, Thousand Oaks, CA
38. Bryman A (2006) Integrating quantitative and qualitative research: how is it done? Qual Res 6:97–113. https://doi.org/10.1177/1468794106058877
39. Bryman A (2012) Social research methods. Oxford University Press
40. Walliman N (2005) Your research project: a step-by-step guide for the first-time researcher. Sage
41. Abrahamson M (1983) Social research methods. Prentice-Hall, Englewood Cliffs, NJ
42. Central Statistical Agency (2007) Abstract and population census reports. CSA: GoE, Addis Ababa, Ethiopia
43. UNDP (2018) Industrialization with a human face. United Nations Development Programme, Addis Ababa, Ethiopia
44. Gebreselassie S, Kirui OK, Mirzabaev A (2016) Economics of land degradation and improvement in Ethiopia. In: Economics of land degradation and improvement—a global assessment for sustainable development. Springer International Publishing, Cham, pp 401–430
45. Capozza DR, Hendershott PH, Mack C (2004) An anatomy of price dynamics in illiquid markets: analysis and evidence from local housing markets. Real Estate Econ 32:1–32. https://doi.org/10.1111/j.1080-8620.2004.00082.x
46. Solow RM, Vickrey WS (1971) Land use in a long narrow city. J Econ Theory 3:430–447
47. Livesey D (1973) Optimum city size: a minimum congestion cost approach. J Econ Theory 6:144–161. https://doi.org/10.1016/0022-0531(73)90031-8

Land Use Change Model Comparison: Mae Sot Special Economic Zone

Sutatip Chavanavesskul and Giuseppe T. Cirella

Abstract Development of the Mae Sot Special Economic Zone (SEZ), Tak province, connects Thailand's economy through the city of Myawaddy, Karen State, Myanmar with Mawlamyine, Yangon, Myanmar, India, and the south of China. Support for several basic infrastructure-related projects and public sector mega department stores are under construction. To date, these investments had not appeared in Tak province. As a result, land use change plays an important part in influencing Mae Sot SEZ. This chapter is a case study on land use change and prediction modeling over the next 20 years (i.e., 2028 and 2038) utilizing the cellular automata (CA)-Markov model and Land Change Modeler (LCM) methods. Predictive results show similar findings from both methods. Results indicate the forest areas and water bodies will change into agricultural and community areas, while the agricultural areas will change to community areas. These methods can assist in proper administrative safe measures to monitor impact on society, environment, security, and public health.

Keywords Urbanization · Land use · Monitoring model · CA-Markov model · Land change modeler · Thailand

1 Introduction

Cooperation within the Association of Southeast Asian Nations has encouraged those countries to establish more than 60 special economic zones (SEZs) specifically focused on investment and financial development [1–3]. These zones specifically expand manufacturing, trade, and service activities, and best practices land usage [2, 4–7]. Mae Sot SEZ, Tak province, as a result, was designated in the East–West economic corridor [8]. It acts as a gateway to the economic center of Yangon, Myanmar

S. Chavanavesskul (✉)
Department of Geography, Srinakharinwirot University, Bangkok, Thailand
e-mail: such2305@gmail.com

G. T. Cirella
Faculty of Economics, University of Gdansk, Sopot, Poland
e-mail: gt.cirella@ug.edu.pl

© Springer Nature Singapore Pte Ltd. 2020
G. T. Cirella (ed.), *Sustainable Human–Nature Relations*,
Advances in 21st Century Human Settlements,
https://doi.org/10.1007/978-981-15-3049-4_7

123

connecting India and southern China through the border of Myawaddy, Karen State, Myanmar. The economic goals of this zone have been to increase competitiveness, employment and enhance the wellness of residents [9, 10]. Once established, this can solve the issues concerning foreign workers, smuggling workers into the interior areas of a country and prevent illegal smuggling of agricultural products from neighboring countries [9–11]. Mae Sot SEZ is the only province receiving investment support from all industry sectors, including road construction at Tanaosri–Kawkareik of section 2–3 of highway no. 12 and the Thai–Myanmar Friendship Bridge II. The expansion of Mae Sot International Airport and mega-investment projects have contributed to several changes to the environment (e.g., drought, flood, and increased land temperature), community well-being (e.g., affecting local peoples' way of life, residential development, electricity consumption, and transportation), and economic success (e.g., tourism) [8–10]. This chapter examines changes in land use within Mae Sot SEZ and forecasts changes by comparing two land use monitoring models: cellular automata-Markov (CA-Markov) model and Land Change Modeler (LCM).

2 Methodology

2.1 Literature Review

The literature review examined three aspects of research focusing in Thailand's SEZs, urbanization processes and land use change modeling. First, SEZs in Thailand are areas with special laws that have been established to promote, facilitate, and provide special privileges in industrial, commercial, and service activities [12]. As an economic hub, these zones promote investment, export to reduce operating cost, advanced technology to produce domestic products, and increase employment. SEZs often focus on industries that require large numbers of workers [8, 12–16]. It has been clearly shown that implementing a SEZ overwhelmingly affects the urbanization of that area [4–6, 17]. It has been shown that SEZ s can have a significant impact on the local economy by increasing foreign direct investment not merely by way of firm relocation but by generating wage increases and infrastructural improvements [10, 18].

Second, urbanization is a process that changes both spatially and socially [19–21]. It incorporates a move from rural to urban regarding spatial and social factors, such as the expansion and improvement of transportation, workplaces and mobility of labor, and housing and utilities. Government policy, economic development, city size, resident demographics, and urban-to-peri-urban fringe development are some of complex holistic-relational linkages being reviewed [13, 22, 23]. Urbanization can be traced using land use change modeling [24]. Given that urbanization is considered a key anthropogenic alteration of the environmental framework, empirical approaches utilizing computed formats such as remote sensing, geographic information systems, and spatiotemporal satellite imagery have become crucial to best observe urban sprawl and urban development practices [25–28]. Land use change in and around urban areas needs proper policy and regulatory management, supported by decision making, that is based on the strong understanding of the urbanization process [27, 29–31]. Monitoring models can be very useful, supportive tools for scenario visualization and optimal localization strategies.

Third, land use monitoring models have been extensively developed by development planners over the last 20 years [32–37]. Currently, there are a number of land use monitoring models used in studying such phenomena by predicting the tendency of change from physical, social, and economic factors [33, 34] as well as, more recently, the urbanization of SEZs which are designated economic development havens. Specific to this chapter, CA-Markov model and LCM are used to study change, predict land use, and comparatively examine model types. Land use models are used to improve and better understand land use change induced by human activities, are important tools for geomatics and environmental research, and monitor and analyze change overtime. The CA-Markov model is considered to be advantageous for modeling land use change due to its robust spatial and temporal dynamic chain analysis approach [38, 39]. LCM, on the other hand, is used mainly to visualize change and produce models that exemplify stable land cover rather than rapidly changing settings [40–42]. LCM can project future plans for a preordained future block of time through historical rates of transition, an important factor that differentiates from the CA-Markov chain analysis.

2.2 Cellular Automata-Markov Model

The CA-Markov model combines cellular automata, Markov chain, multi-criteria, and multi-objective land allocation to predict land use change over time [12, 26, 27, 43–45]. Moreover, it analyzes land use change spatially as well as temporally [44]. The CA-Markov model can be used to analyze several kinds of land use variables and patterns [38, 39]. It is suitable for monitoring land use and land cover change within the scope of creating raster model applications [43, 44]. It requires two temporal datasets to create the probability of change from metric form to be able to predict future land use change [45–50] (Eq. 1).

$$V_j P_{jk} = [V_1, V_2, V_3, \ldots V_m]_2 \begin{bmatrix} P_{1,1} & P_{1,2} & \ldots & P_{1,m\ldots} \\ P_{2,1} & P_{2,2} & \ldots & P_{2,m\ldots} \\ P_{m,1} & P_{m,2} & \ldots & P_{m,m\ldots} \end{bmatrix}$$
$$= [V_1, V_2, V_3, \ldots V_m]_3 \tag{1}$$

where $V_j P_{jk}$ = future land use (i.e., V_j = annual land use and P_{jk} = probability of changing at the time 'j' and 'k'). However, it is worth noting that human decision making is another factor, regardless of CA-Markov model results, since policy is often determined at a local scale [24, 43, 45].

2.3 Land Change Modeler

As an innovative land planning and decision support software tool, LCM is a hybrid-based model developed to optimize the efficiency of land use simulation based on two temporal land use data comparisons: simulation and probability of each land use type [51, 52]. Apart from physical and temporal factors [51], it can be applied to determine demographic and social factors derived from human decision makings (e.g., government policy, development planning of the private sector, and the community at large) [41]. Generally, population and social factors are interlaced and often affect such land use changes [53]. In term of analytical output, it is performed using probability metrics for each land use type [14]. LCM analyzes relationships via Cramér's V measurement in which if the value is greater than 0.15 factors, then it is considered to have influenced change [54]. Moreover, LCM typically uses multilayer perceptron, creating potential areas for each type of land use change [14, 55, 56]. Utilizing this method, evidence to explore a cause-and-effect relationship of urbanization and land use change, from present to future, is critical in investigating the experimentation of the Mae Sot SEZ.

2.4 Study Process

The study process is broken down into three parts: exploratory, preparatory, and temporal. First, the study area, Mae Sot SEZ, covers 128 km^2 in Tambon Mae Sot and some parts of Mae Pa, Tambon Tha Sai Luad, Tambon Mae Taw, Tambon Phra That Pha Daeng, Amphoe Tak, including the connected area between Mae Sot border and Myawaddy, Myanmar (Fig. 1).

Second, prepared Landsat 5 satellite images are utilized. These images are adjusted to integrate spectral and geometric correction to reduce ambiguity from atmospheric distortion and any incorrect signaling [42]. The images were rectified, corresponding with the geographical coordinate system and enhanced to sharpen the distinction of objects on the images [57]. The land use classification is categorized into five classes (i.e., types of area): (1) forest, (2) agriculture, (3) community, (4) water, and (5) miscellaneous land. Six field surveys validated each land use class, totaling 30 field surveys with average of 80% accuracy. The most accurate data is the agriculture and community area (i.e., 100% accurate), while the least is the forest area (i.e., 50% accurate) (Table 1).

Third, temporal land use data was selected over a 30-year period in 1988, 1998, 2008, and 2018. Almost all temporal data were defined before the declaration of Mae Sot SEZ, with exception to 2018. The predictions of land use change duration are made in 10-year increments (i.e., 2028 and 2038) by using the land use monitoring models.

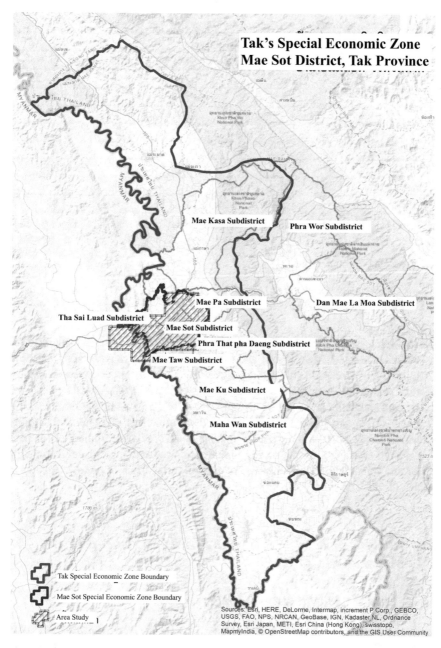

Fig. 1 Study area

Table 1 Accuracy of field surveys regarding the categorized land use classes

Land use[a]	Survey points					
	F	A	C	W	M	Total
F	3	–	–	–	–	3
A	2	6	–	1	2	11
C	1	–	6	–	–	7
W	–	–	–	5	–	5
M	–	–	–	–	4	4
Total	6	6	6	6	6	30
Correspond	3	6	6	5	4	–
Accuracy (%)	50	100	100	83.33	66.67	80

[a]F forest, A agriculture, C community, W water, M miscellaneous land

3 Results

3.1 Land Use Change

Clearly, land use in Mae Sot SEZ has changed incrementally over the past three decades (i.e., 1988–2018). Forest and agriculture have decreased continuously in contrast to the increase in community, miscellaneous land, and water areas. Annual results indicate that community areas have increased by 9.12% followed by miscellaneous land and water areas (i.e., 2.63 and 2.35%), while forest and agriculture have declined 3.01% and 0.24%, respectively. Urbanization has expanded and rapidly grown since late 2010 with the policy implementation and development of Mae Sot SEZ. Major expansion from old communities to newer ones has appeared along the major and minor transportation lines, in particular, the Thai–Myanmar Friendship Bridge (Table 2 and Fig. 2).

3.2 Analysis of Land Use Change Prediction by CA-Markov Model and LCM

3.2.1 Land Use Prediction by CA-Markov Model

Predicting land use for 2028 and 2038 is based on land use data from 2008 and 2018. Using the CA-Markov model, the results were estimated to be 82.76% accurate. In 2028, the results of the analysis are the proportion of change (Table 3) and probability of change (Table 4) from the CA-Markov model. Results for the next decade (i.e., 2038) follow suite (Tables 5 and 6).

It was found in 2018 that the major areas of Mae Sot SEZ were agriculture (i.e., 51.93%), followed by community (i.e., 42.59%), miscellaneous land (i.e., 3.21%),

Table 2 Land use of Mae Sot SEZ from 1988 to 2018, using Landsat 5 images

Land use[a]	Mae Sot SEZ					
	1988	1998	2008	2018	Change from 1988 to 2018	Annual change rate (%)
F	30.70 (23.98%)	21.04 (16.43%)	7.07 (5.52%)	3.02 (2.36%)	(−)27.68 (−21.63%)	(−)3.01
A	82.19 (64.21%)	85.12 (66.50%)	86.48 (67.56%)	76.39 (59.68%)	(−)5.8 (−4.53%)	(−)0.24
C	11.12 (8.69%)	17.13 (13.38%)	28.67 (22.40%)	41.55 (32.46%)	30.43 (23.77%)	9.12
W	1.16 (0.90%)	1.38 (1.08%)	1.95 (1.53%)	1.97 (1.54%)	0.82 (0.64%)	2.35
M	2.83 (2.21%)	3.33 (2.60%)	3.83 (2.99%)	5.06 (3.96%)	2.24 (1.75%)	2.63
Total	128.00 (100%)	128.00 (100%)	128.00 (100%)	128.00 (100%)		

[a]*F* forest, *A* agriculture (i.e., paddy fields, perennials, fruit orchards, and horticultural crops), *C* community (i.e., towns and commercial areas, residences, transportation stations, other buildings, and golf courses), *W* water, *M* miscellaneous land (i.e., meadows and woodlands, marshes and swamps, mines, industrial ponds, and other miscellaneous areas)

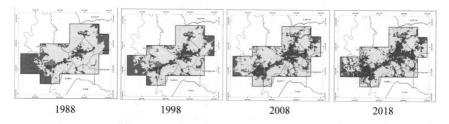

1988　　　　1998　　　　2008　　　　2018

Fig. 2 Land use map of Mae Sot SEZ from 1988 to 2018, using Landsat 5 images; Area colors from Table 2 are overlaid

Table 3 Proportion of land use change in 2028 from the CA-Markov model

Land use[a]	Proportion of land use change in 2028				
	F	A	C	W	M
F	0.9693	0.6957	0.3951	–	0.936
A	–	64.026	10.5498	0.0972	1.7343
C	–	–	41.544	0.0054	–
W	–	0.0396	0.0243	1.8369	0.0594
M	–	1.6137	2.0043	0.0531	1.377

[a]*F* forest, *A* agriculture, *C* community, *W* water, *M* miscellaneous land

Table 4 Probability of land use change in 2028 from the CA-Markov model

Land use[a]	Proportion of land use change in 2028				
	F	A	C	W	M
F	0.3237	0.2323	0.1318	–	0.3123
A	–	0.8380	0.1381	0.0013	0.0227
C	–	–	0.9999	0.0001	–
W	–	0.0200	0.0124	0.9374	0.0302
M	–	0.3196	0.3971	0.0105	0.2728

[a]F forest, A agriculture, C community, W water, M miscellaneous land

Table 5 Proportion of land use change in 2038 from the CA-Markov model

Land use[a]	Proportion of land use change in 2038				
	F	A	C	W	M
F	0.1404	1.6893	0.8406	0.0135	0.3123
A	0.135	56.6901	17.3871	0.5742	1.6209
C	–	–	41.5494	–	–
W	–	0.0468	0.0279	1.8054	0.0783
M	0.0018	2.3202	2.0448	0.0630	0.6183

[a]F forest, A agriculture, C community, W water, M miscellaneous land

Table 6 Probability of land use change in 2038 from the CA-Markov model

Land use[a]	Proportion of land use change in 2038				
	F	A	C	W	M
F	0.0468	0.5637	0.2807	0.0044	0.1043
A	0.0018	0.7419	0.2276	0.0075	0.0212
C	–	–	1.0000	–	–
W	–	0.0241	0.0141	0.9216	0.0402
M	0.0004	0.4596	0.405	0.0125	0.1225

[a]F forest, A agriculture, C community, W water, M miscellaneous land

water (i.e., 1.51%), and forest (i.e., 0.76%), respectively. Comparatively, in 2038, the major areas change with community (i.e., 48.34%), followed by agriculture (i.e., 47.49%), miscellaneous land (i.e., 2.03%), water (i.e., 1.92%), and forest (i.e., 0.22%), respectively. The highest increasing proportion is the water body which increases 1.37% annually, followed by community (i.e., 0.67%), while the forest area decreases by 3.54% annually, followed by miscellaneous land (i.e., 1.84%) and agriculture (i.e., 0.43%). Remarkably, communities will continually expand along the main transportation routes and around the Thai–Myanmar Friendship Bridge (Table 7 and Fig. 3).

Table 7 Land use of Mae Sot SEZ between 2028 and 2038 using the CA-Markov model

Land use[a]	Mae Sot SEZ			
	2028	2038	Changes from 2028 to 2038	Yearly change rates (%)
F	0.97 (0.76%)	0.28 (0.22%)	(−)0.69 (−0.54%)	(−)3.54
A	66.47 (51.93%)	60.79 (47.49%)	(−)5.66 (−4.44%)	(−)0.43
C	54.52 (42.59%)	61.87 (48.34%)	7.35 (5.74%)	0.67
W	1.93 (1.51%)	2.46 (1.92%)	0.53 (0.41%)	1.37
M	4.11 (3.21%)	2.60 (2.03%)	(−)1.51 (−1.18%)	(−)1.84
Total	128.00 (100%)	128.00 (100%)		

[a]*F* forest, *A* agriculture (i.e., paddy fields, perennials, fruit orchards, and horticultural crops), *C* community (i.e., towns and commercial areas, residences, transportation stations, other buildings, and golf courses), *W* water, *M* miscellaneous land (i.e., meadows and woodlands, marshes and swamps, mines, industrial ponds, and other miscellaneous areas)

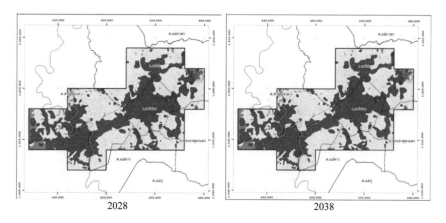

Fig. 3 Land use map of Mae Sot SEZ between 2038 and 2038 by CA-Markov model, using Landsat 5 images; Area colors from Table 7 are overlaid

3.2.2 Land Use Prediction by LCM

Using the same starting points as with the CA-Markov model, LCM predicted land use change for 2028 and 2038 based 2008 and 2018 data. It also considered physical factors, such as digital elevation modeling, slope, aspect, village distance, road distance, stream distance, and social factors (i.e., government policy and population density). The LCM analysis results were an average of 83.91% accurate. In 2028, the results of the analysis are the proportion of change (Table 8) and probability of change (Table 9) from LCM, and following this, supplemental 2038 results are presented (Tables 10 and 11).

Inference from the LCM in 2028 revealed that the major areas of Mae Sot SEZ will be agriculture (i.e., 51.88%), followed by community (i.e., 42.59%), miscellaneous land (i.e., 3.21%), water (i.e., 1.56%), and forest (i.e., 0.76%), respectively. For the decade that follows in 2038, the major areas will be agriculture (i.e., 48.32%), followed by miscellaneous land (i.e., 2.06%), water (i.e., 1.92%), and forest (i.e., 0.22%), respectively.

The highest proportion of increased area is the water body which will increase 1.16% annually followed by the community area (i.e., 0.67%), while forest, miscellaneous land, and agriculture will decrease (i.e., 3.57%, 1.80%, and 0.42%, respectively). Analogous with the CA-Markov model, the expansion of communities will continually expand along the main transportation routes and around the Thai–Myanmar Friendship Bridge (Table 12 and Fig. 4).

Table 8 Proportion of land use change in 2028 from LCM

Land use[a]	Proportion of land use change in 2028				
	F	A	C	W	M
F	0.9702	0.6957	0.3951	–	0.936
A	–	64.0296	10.5516	0.099	1.7343
C	–	–	41.5449	0.0045	–
W	–	0.0396	0.0243	1.8369	0.0594
M	–	1.6137	2.0043	0.0531	1.377

[a]F forest, A agriculture, C community, W water, M miscellaneous land

Table 9 Probability of land use change in 2028 from LCM

Land use[a]	Proportion of land use change in 2028				
	F	A	C	W	M
F	0.3237	0.2323	0.1318	–	0.3123
A	–	0.8380	0.1381	0.0013	0.0227
C	–	–	0.9999	0.0001	–
W	–	0.0200	0.0124	0.9374	0.0302
M	–	0.3196	0.3971	0.0105	0.2728

[a]F forest, A agriculture, C community, W water, M miscellaneous land

Table 10 Proportion of land use change in 2038 from LCM

Land use[a]	Proportion of land use change in 2038				
	F	A	C	W	M
F	0.1404	1.6893	0.8406	0.0135	0.3123
A	0.1377	56.6865	17.3907	0.5733	1.6200
C	–	–	41.5494	–	–
W	–	0.0468	0.0279	1.8054	0.0792
M	–	2.3202	2.0448	0.063	0.6201

[a]F forest, A agriculture, C community, W water, M miscellaneous land

Table 11 Probability of land use change in 2038 from LCM

Land use[a]	Proportion of land use change in 2038				
	F	A	C	W	M
F	0.0468	0.5637	0.2807	0.0044	0.1043
A	0.0018	0.7419	0.2276	0.0075	0.0212
C	–	–	1.0000	–	–
W	–	0.0241	0.0141	0.9216	0.0402
M	0.0004	0.4596	0.405	0.0125	0.1225

[a]F forest, A agriculture, C community, W water, M miscellaneous land

Table 12 Land use of Mae Sot SEZ from 2028 to 2038 by LCM

Land use[a]	Mae Sot SEZ			
	2028	2038	Changes from 2028 to 2038	Yearly change rates (%)
F	0.97 (0.76%)	0.28 (0.22%)	(−)0.69 (−0.54%)	(−)3.57
A	66.41 (51.88%)	60.78 (47.48%)	(−)5.63 (−4.40%)	(−)0.42
C	54.52 (42.59%)	61.85 (48.32%)	7.33 (5.73%)	0.67
W	1.99 (1.56%)	2.46 (1.92%)	0.46 (0.36%)	1.16
M	4.11 (3.21%)	2.63 (2.06%)	(−)1.48 (−1.15%)	(−)1.80
Total	128.00 (100%)	128.00 (100%)		

[a]F forest, A agriculture (i.e., paddy fields, perennials, fruit orchards, and horticultural crops), C community (i.e., towns and commercial areas, residences, transportation stations, other buildings, and golf courses), W water, M miscellaneous land (i.e., meadows and woodlands, marshes and swamps, mines, industrial ponds, and other miscellaneous areas)

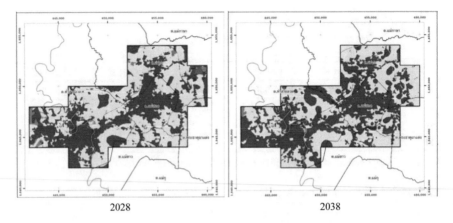

2028   2038

Fig. 4 Land use map of Mae Sot SEZ from 2038 to 2038 by LCM, using Landsat 5 images; Area colors from Table 12 are overlaid

3.3 Comparative Results

Regarding the predictive results, based on the land use data from 2008 and 2018, using the CA-Markov model and LCM, land use change in 2028 and 2038 showed very similar to almost identical forecasting for each year. Generally, the forest area and water bodies will change to agriculture and community areas. Extraordinarily, these models showed the differentiation of spatial data of each land use type to the degree of less than 1 km^2. It was found that the spatial differentiation of both models in 2028 is higher than those in 2038. Considering this effect, agriculture and water bodies differ from around 0.06 km^2 in 2028. In contrast, in 2038, a decade later, the models indicate greater differentiation in agriculture, community, and miscellaneous land areas (Table 13). These future-base comparative results means The predicted results can be useful in developing policy needs and proper administrative safe measures.

4 Conclusion

The decision to select a method of classification for land use change that appropriates the characteristics of the image is essential. The use of the CA-Markov model and LCM fits our object-based land use classifications [12, 26, 43]. The strengths of these methods are that they show high levels of accuracy (i.e., especially in terms of classification results). The two methods did not differ in predicting land use change in Mae Sot SEZ. This was found, even with the introduction of two social factors added to the LCM analysis. It should be noted, the size of analyzed area for the two models did not differ, and the resolution was considered moderate; however, increasing the resolution would refine more detailed results. It was shown, with the addition

Table 13 Comparing land use change between the CA-Markov model and LCM

Land use[a]	Mae Sot SEZ					
	CA-Markov		LCM		Difference between CA-Markov and LCM	
	2028	2038	2028	2038	2028	2038
F	0.97 (0.76%)	0.28 (0.22%)	0.97 (0.76%)	0.28 (0.22%)	— –	— –
A	66.47 (51.93%)	60.79 (47.49%)	66.41 (51.88%)	60.78 (47.48%)	0.06 (0.05%)	0.01 (0.01%)
C	54.52 (42.59%)	61.87 (48.34%)	54.52 (42.59%)	61.85 (48.32%)	— —	0.02 (0.02%)
W	1.93 (1.51%)	2.46 (1.92%)	1.99 (1.56%)	2.46 (1.92%)	0.06 (0.05%)	— —
M	4.11 (3.21%)	2.60 (2.03%)	4.11 (3.21%)	2.63 (2.06%)	— –	0.03 (0.03%)
Total	128.00 (100%)	128.00 (100%)	128.00 (100%)	128.00 (100%)		

[a]F forest, A agriculture, C community, W water, M miscellaneous land

of social factors in the LCM, insufficient spatial variation could be observed. A key importance to furthering the study would be the development of other potential factors that influence land use change such as disaster risk, government policy, and other economic indicators. Likewise, a more extensive study area and higher resolution land use classification would supplement future research.

It can be concluded that the Mae Sot SEZ caused some developmental strife between Thailand and Myanmar [9, 17]. Discord has been linked to uneven economic development, tourism and travel, and trade. As with all land-specific agreements, advantages and disadvantages need to be weighed and considered. As such, surrounding communities with the SEZ have noted enhanced access to education, economic opportunity, travel, infrastructure, basic utilities, and transportation. These developments, to proximate communities, play an important part in augmenting poverty as well as livelihood. On the downside, concerns with drastic change in land use, land price bidding, isolating locals due to price increases, and transformation of land ownership need to be addressed [8]. Additional land use and land management policies related to SEZs should closely examine legal mechanisms that overlook the level of participation and long-term development outcomes. Foreign investment-related policy needs to be prioritized, especially in terms of access to land resources and eligible activities. One potential idea is the allocation of a range of incentives to investors willing to provide capital. Labor policy within Thailand's SEZs needs to be managed as returning and seasonal cross-border foreign workers can become problematic if left unattended. Finally, due to the influx of persons, properly administrative safe measures should monitor the impacts on society, environment, security, and public health. Land use change models provide an essential observational tool for these measures and governmental assistance in coordinating future planning and sustainability-oriented development.

Acknowledgements Gratitude and funding are given to Srinakharinwirot University, Bangkok, Thailand.

References

1. Tsertseil J (2015) The clusters and special economic zone: the improvement in the development of the region. OMICS International
2. Frick SA, Rodríguez-Pose A, Wong MD (2019) Toward economically dynamic special economic zones in emerging countries. Econ Geogr 95:30–64. https://doi.org/10.1080/00130095. 2018.1467732
3. Crane B, Albrecht C, Duffin KM, Albrecht C (2018) China's special economic zones: an analysis of policy to reduce regional disparities. Reg Stud Reg Sci 5:98–107. https://doi.org/ 10.1080/21681376.2018.1430612
4. Babita M (2017) Output and input efficiency of special economic zones (SEZs) in India. Indian Econ J 65:107–118. https://doi.org/10.1177/0019466217727881
5. Wang J (2013) The economic impact of special economic zones: evidence from Chinese municipalities. J Dev Econ 101:133–147
6. Sinenko O, Mayburov I (2017) Comparative analysis of the effectiveness of special economic zones and their influence on the development of territories. Int J Econ Financ Issues 7:115–122
7. Cardinale M, Brusetti L, Quatrini P et al (2004) Comparison of different primer sets for use in automated ribosomal intergenic spacer analysis of complex bacterial communities. Appl Environ Microbiol 70:6147–6156
8. Chavanavesskul S (2018) The impact of Mae Sot's special economic zone on climate change. In: 2018 Asia Global land programme conference: Transitioning to sustainable development of land systems through teleconections and telecoupling, 3–5 Sept 2018. Taipei, Taiwan
9. BOI (2018) A guide to investment in the special economic development zones (SEZ). Office of the Board of Investment, Bangkok
10. Ishida M (2009) Special economic zones and economic corridors. In: Kuchiki A, Uchikawa S (eds) Research on development strategies for CLMV countries. ERIA, Jakarta, pp 33–52
11. Government of Thailand (2014) Results of the special economic development board (NRP) meeting. Government House, Bangkok, Thailand
12. Thitawadee S, Yoshihisa M (2018) Urban growth prediction of special economic development zone in Mae Sot District, Thailand. Eng J 22:269–277. https://doi.org/10.4186/ej.2018.22. 3.269
13. Tawilpipatkul D (1996) The process of urbanisation and social change. Chulalongkorn University, Bangkok, Thailand
14. Labs Clark (2013) IDRISI spotlight: the land change modeler. Clark University, Massachusetts
15. Lee N (2019) Inclusive growth in cities: a sympathetic critique. Reg Stud 53:424–434. https:// doi.org/10.1080/00343404.2018.1476753
16. McNevin A (2007) Irregular migrants, neoliberal geographies and spatial frontiers of 'the political'. Rev Int Stud 33:655–674. https://doi.org/10.1017/S0260210507007711
17. Intarat P (2018) From SEZs to Thailand 4.0: geopolitics of borderlands in the Thai state's vision. For Soc 2:65. https://doi.org/10.24259/fs.v2i1.3600

18. Joshi R (2017) Assessing the impact of income inequality on economic growth. Indian Econ J 65:1–26. https://doi.org/10.1177/0019466217727811
19. De Jong M, Joss S, Schraven D et al (2015) Sustainable-smart-resilient-low carbon-eco-knowledge cities; making sense of a multitude of concepts promoting sustainable urbanization. J Clean Prod 109:25–38. https://doi.org/10.1016/j.jclepro.2015.02.004
20. UN (2018) 2018 revision of world urbanization prospects. New York
21. Russo A, Cirella G (2018) Modern compact cities: how much greenery do we need? Int J Environ Res Public Health 15:2180. https://doi.org/10.3390/ijerph15102180
22. Long H, Heilig GK, Li X, Zhang M (2007) Socio-economic development and land-use change: analysis of rural housing land transition in the transect of the Yangtse River, China. Land Use Policy 24:141–153. https://doi.org/10.1016/j.landusepol.2005.11.003
23. International Institute for Trade and Development (2014) Guidelines and measures for the development of special economic zones in Thai border. Policy briefs on research projects: IITD, Bangkok, Thailand
24. Losiri C (2017) Land use change model and urban area prediction in the future. J Soc Sci 19:340–357
25. Oueslati W, Salanié J, Wu J (2019) Urbanization and agricultural productivity: some lessons from European cities. J Econ Geogr 19:225–249. https://doi.org/10.1093/jeg/lby001
26. Mosammam HM, Nia JT, Khani H et al (2017) Monitoring land use change and measuring urban sprawl based on its spatial forms: the case of Qom city. Egypt J Remote Sens Sp Sci 20:103–116. https://doi.org/10.1016/j.ejrs.2016.08.002
27. Deep S, Saklani A (2014) Urban sprawl modeling using cellular automata. Egypt J Remote Sens Sp Sci 17:179–187. https://doi.org/10.1016/j.ejrs.2014.07.001
28. Fuglsang M, Münier B, Hansen HS (2013) Modelling land-use effects of future urbanization using cellular automata: an Eastern Danish case. Environ Model Softw 50:1–11. https://doi.org/10.1016/J.ENVSOFT.2013.08.003
29. Thorp KR, Bronson KF (2013) A model-independent open-source geospatial tool for managing point-based environmental model simulations at multiple spatial locations. Environ Model Softw 50:25–36. https://doi.org/10.1016/j.envsoft.2013.09.002
30. Patra S, Sahoo S, Mishra P, Mahapatra SC (2018) Impacts of urbanization on land use cover changes and its probable implications on local climate and groundwater level. J Urban Manag 7:70–84. https://doi.org/10.1016/J.JUM.2018.04.006
31. Verburg PH, Schot PP, Dijst MJ, Veldkamp A (2004) Land use change modelling: current practice and research priorities. GeoJournal 61:309–324. https://doi.org/10.1007/s10708-004-4946-y
32. Terama E, Clarke E, Rounsevell MDA et al (2019) Modelling population structure in the context of urban land use change in Europe. Reg Environ Chang 19:667–677. https://doi.org/10.1007/s10113-017-1194-5
33. Cao Q, Yu D, Georgescu M et al (2018) Impacts of future urban expansion on summer climate and heat-related human health in eastern China. Environ Int 112:134–146. https://doi.org/10.1016/j.envint.2017.12.027
34. Yang B, Yang X, Leung LR et al (2019) Modeling the impacts of urbanization on summer thermal comfort: the role of urban land use and anthropogenic heat. J Geophys Res Atmos 124:2018JD029829. https://doi.org/10.1029/2018JD029829
35. Buzan JR, Oleson K, Huber M (2015) Implementation and comparison of a suite of heat stress metrics within the community land model version 4.5. Geosci Model Dev 8:151–170. https://doi.org/10.5194/gmd-8-151-2015
36. Long H, Wu X, Wang W, Dong G (2008) Analysis of urban-rural land-use change during 1995–2006 and its policy dimensional driving forces in Chongqing, China. Sensors 8:681–699. https://doi.org/10.3390/s8020681
37. Yao X, Wang Z, Wang H (2015) Impact of urbanization and land-use change on surface climate in middle and lower reaches of the Yangtze River, 1988–2008. Adv Meteorol 2015:1–10. https://doi.org/10.1155/2015/395094

38. Wang SQ, Zheng XQ, Zang XB (2012) Accuracy assessments of land use change simulation based on Markov-cellular automata model. Procedia Environ Sci 13:1238–1245. https://doi.org/10.1016/j.proenv.2012.01.117

39. Halmy MWA, Gessler PE, Hicke JA, Salem BB (2015) Land use/land cover change detection and prediction in the north-western coastal desert of Egypt using Markov-CA. Appl Geogr 63:101–112. https://doi.org/10.1016/J.APGEOG.2015.06.015

40. Noszczyk T (2019) A review of approaches to land use changes modeling. Hum Ecol Risk Assess An Int J 25:1377–1405. https://doi.org/10.1080/10807039.2018.1468994

41. Mishra V, Rai P, Mohan K (2014) Prediction of land use changes based on land change modeler (LCM) using remote sensing: a case study of Muzaffarpur (Bihar), India. J Geogr Inst Jovan Cvijic, SASA 64:111–127. https://doi.org/10.2298/IJGI1401111M

42. Roshanbakhsh S, Modaresi SA, Karami J (2017) Land use changes using multi-layer perception and change modeler. Int J Urban Manag Energy Sustain 1:79–84. https://doi.org/10.22034/IJUMES.2017.01.01.008

43. Hyandye C, Martz LW (2017) A Markovian and cellular automata land-use change predictive model of the Usangu Catchment. Int J Remote Sens 38:64–81. https://doi.org/10.1080/01431161.2016.1259675

44. Ghosh P, Mukhopadhyay A, Chanda A et al (2017) Application of cellular automata and Markov-chain model in geospatial environmental modeling—a review. Remote Sens Appl Soc Environ 5:64–77. https://doi.org/10.1016/J.RSASE.2017.01.005

45. Kumar S, Radhakrishnan N, Mathew S (2014) Land use change modelling using a Markov model and remote sensing. Geomatics, Nat Hazards Risk 5:145–156. https://doi.org/10.1080/19475705.2013.795502

46. Mondal MS, Sharma N, Garg PK, Kappas M (2016) Statistical independence test and validation of CA Markov land use land cover (LULC) prediction results. Egypt J Remote Sens Sp Sci 19:259–272. https://doi.org/10.1016/J.EJRS.2016.08.001

47. Hishe S, Bewket W, Nyssen J, Lyimo J (2019) Analysing past land use land cover change and CA-Markov-based future modelling in the Middle Suluh Valley, Northern Ethiopia. Geocarto Int 1–31. https://doi.org/10.1080/10106049.2018.1516241

48. Liping C, Yujun S, Saeed S (2018) Monitoring and predicting land use and land cover changes using remote sensing and GIS techniques-a case study of a hilly area, Jiangle, China. PLoS One 13:e0200493. https://doi.org/10.1371/journal.pone.0200493

49. Nouri J, Gharagozlou A, Arjmandi R et al (2014) Predicting urban land use changes using a CA–Markov model. Arab J Sci Eng 39:5565–5573. https://doi.org/10.1007/s13369-014-1119-2

50. Hamad R, Balzter H, Kolo K (2018) Predicting land use-land cover changes using a CA-Markov model under two different scenarios. Sustainability 10:3421. https://doi.org/10.3390/su10103421

51. Russo P (2017) Usability of planning support systems: analysing adoption and use in planning practice. University of Melbourne

52. Joorabian Shooshtari S, Shayesteh K, Gholamalifard M et al (2017) Impacts of future land cover and climate change on the water balance in northern Iran. Hydrol Sci J 62:2655–2673. https://doi.org/10.1080/02626667.2017.1403028

53. Widyasamratri H, Aswad A (2017) A preliminary study: an agent-based spatial simulation of human-coastal environment interaction. In: The third international conference on coastal and delta areas, pp 593–601

54. Maestripieri N (2012) Dynamiques spatio-temporelles des plantationsforestières industrielles dans le sud chilien: de l'analysediachronique à la modélisation prospective [in French]. Université Toulousele Mirail, Toulouse II

55. Clark Labs (2019) Clark labs. https://clarklabs.org/. Accessed 26 May 2019

56. Nadnicha P, Onanong P, Kasem C et al (2016) The effect of land use changes on landslide in the high slope area at Surat Thani Province. KKU Sci J 44:212–221

57. Osman TMK (2016) Driving Forces, and future directions of informal urban expansion in greater Cairo Metropolitan region. Kyushu University

Cohesion Policy for Europe 2020

Andrzej Paczoski, Solomon T. Abebe, and Giuseppe T. Cirella

Abstract Cohesion policy is one of the more important elements of the European Union's (EU) integrated processes. Evaluating this type of mechanism can show the efficiency of European policy convergence and its elimination of development inequalities between Member States (MSs). The Europe 2020 Strategy introduced new challenges for MSs through strategic goals piloted by targets. MSs work at creating opportunities in an endogenous Union in which cohesion policy is formulated to best conflate and balance cooperation. This chapter examines cohesion policy as a tool for harmonized development by enlightening what are the, and how to decrease, development disparities among MSs. Europe 2020 targets are examined from 2011 to 2018. Overall, best results were found in Scandinavia, Benelux, and Northern Western countries, while Southern countries recorded low results. Specific MS-concerns, highlighted at the national-level, include lack of effective institutional framework, transport infrastructure, education, and innovation policy.

Keywords Cohesion reporting · Europe 2020 Strategy · Doing business · EU

1 Introduction

In the twenty-first century, new challenges have veered the European Union (EU) toward increasing its global competitiveness. The financial crisis since 2008 negatively influenced Europe, obliging additional responsibilities to Member States' (MSs) willingness to strengthen the Union's economy and harmonize development in a manner that decreases inequality and level of underdevelopment. Cohesion policy

A. Paczoski (✉) · G. T. Cirella
Faculty of Economics, University of Gdansk, Sopot, Poland
e-mail: andrzej.paczoski@ug.edu.pl

G. T. Cirella
e-mail: gt.cirella@ug.edu.pl

S. T. Abebe
Polo Centre of Sustainability, Imperia, Italy
e-mail: solomtu6@gmail.com

© Springer Nature Singapore Pte Ltd. 2020
G. T. Cirella (ed.), *Sustainable Human–Nature Relations*,
Advances in 21st Century Human Settlements,
https://doi.org/10.1007/978-981-15-3049-4_8

is one of the more important elements in the integration processes of the EU. MSs are responsible for supporting and pursuing opportunities in endogenous growth through mechanisms that assist in formulating robust, sound cohesion policy. This chapter examines the use of cohesion policy as a tool for harmonizing development by looking at interconnected differences and how to decrease development disparities among MSs. A key premise to the development of this chapter is how EU policies evolved over the last few decades and whether cohesion policy was effectively implemented for the Europe 2020 Strategy. An evaluation of this mechanism can show the efficiency of cohesion policy and potential convergence processes as well as the elimination of development inequalities among MSs. Molle [1] stated that both theoretical and practical research in a clear and accessible "structure, covering economic, social, and territorial issues, […] provides a systematic view of the various stages of the whole policy cycle." An examination of the policy cycle, assessing the problems and identifying their causes, is the core focus of this research. The Europe 2020 targets are examined from 2011 to 2018 in which the essence of the Europe 2020 Strategy is connected. This chapter elucidates on key concepts of cohesion policy, an extensive look at the 2020 targets, and a discussion on the state-of-the-art for businesses and future outlook in regard to policy effectiveness and consistency.

2 Understanding Cohesion Policy

The term cohesion policy first appeared in the 1957 Rome Treaty which referenced the strengthening of Europe's economy and harmonization of its development. The concept encouraged a decrease in inequality and level of underdevelopment among Europe's regions and countries [2, 3]. With the process of promoting peace, developing economic trade, and co-supporting political institutions among MSs, the EU was formed. Since its inception, on February 7, 1992, the Maastricht Treaty, officially the Treaty on the European Union, was signed in prospect of creating equal levels of inclusivity structuring much of the Union's enlargement and continuing to aspire candidate countries to join [4]. To best understand cohesion policy, one must theorize the integration and conceptualization of policy as a cyclic and evolving process—inclusive of economic, social, and territorial cohesion-oriented action [5]. Reducing these disparities among MSs will also reduce membership hierarchy and separation between the most important with the least favored. Important factors for cohesive integrity will need to focus on rural development, areas affected by industrial transition, and areas which suffer from severe and permanent natural or demographic handicaps (e.g., Europe's northernmost areas with low population density as well as its islands, its cross-border areas, and mountain regions). Cohesion policy is a tool intended to help correct regional imbalance through participation in co-development and structural adjustments [6]. It is emphasized as one of the most important directives in European policy [7]. To best elucidate the state-of-the-art, an examination into post-2010 reporting, a review of the 2020 targets and a brief look at the financial resources dictating cohesion policy are probed.

2.1 Cohesion Reporting Since 2010

The European Commission puts together a cohesion report every three which states the progress of achieving economic and social cohesion across the EU. Since 2010, the last three cohesion reports (i.e., the fifth in 2010, the sixth in 2014, and the seventh in 2017), reporting affirmed important socioeconomic impact and contributive policies from MSs as well as activities associated with intergovernmental institutions such as the European Central Bank. A brief examination of these reports is highlighted to form a state-of-the-art.

The fifth cohesion report, published at the end of 2010, was produced after the 2008 world financial and economic crisis. MSs' attempt to avoid the effects of the crisis was implemented in several ways. Particular attention was focused on keeping enterprises operative, maintaining employment, stimulating economic active (i.e., through demand), and increasing overall public investment. Southern countries (i.e., Greece, Italy, Spain, and Portugal) and Ireland had enormous difficulty with general government debt and deficit as well as a significant decrease in gross domestic product (GDP) [8]. Likewise, several other countries had troubles with general government debt and central budgetary issues that cause a decrease in revenue and increase in expenditure on social services.

This report mainly focused on supporting the development of the Europe 2020 Strategy in conjunction with the Lisbon Strategy. These two documents emphasized innovation, employment, social inclusion, environmental challenges, and climate change. The report highlighted a cooperative Europe 2020 Strategy overlap where cohesion and other EU policies were to work in unison for more effectiveness [9]. Interestingly, the report introduced new factors for analyzing regional economic disparities using measurements, including effectiveness of institutions, index of competitiveness, and indicators of well-being. The following topics make up the four chapters of the report: (1) economic, social, and territorial situation and trends (i.e., to promote economic competitiveness and convergence, improve well-being with the reduction of social exclusion, and enhance environmental sustainability), (2) assess the contribution of national policies to cohesion, (3) establish an overview on how other EU policies contribute to cohesion, and (4) summarize evidence on strengths of cohesion policy in furthering cohesion objectives as well as highlighting issues for its improvement [8].

In regards to the Lisbon Strategy, the fifth cohesion report adopted a number of interconnecting territorial ideas, including joint economic and social improvement, climate change-related strategies and environmentally friendly reform, and how to best measure territorial impact on implemented policy. Moreover, since the Lisbon Strategy was framed, the key connection of overcoming the differences in growth and productivity among MSs and the world's leading competitors (i.e., the USA, China, and Japan) became an ongoing dynamic and competitive contest. A knowledge-based economy capable of sustainable economic growth, by creating more and better jobs and greater social cohesion with environmental awareness, is at the core of the strategy [10]. The strategy is based on the three pillars of sustainability (i.e., ecological, social, and economic) and accenting to technological capacity and innovation as the way forward for European competitiveness.

The European Commission relaunched the Lisbon Strategy in 2005 due to initial weak results. The revised Lisbon Strategy was further based on three priorities: (1) creating a more attractive environment for investment and work (e.g., expansion of the common market, improvement of European and national law, and creation of open and competitive markets), (2) improving knowledge and innovation for economic growth (i.e., more outlays on research and technological development (RTD), innovation, and sustainable use of resources), and (3) creating more and better jobs (i.e., modernizing the system of social security, engaging more people in employment, and increasing the ability to adapt for market requirements) [10]. The fifth cohesion report was formulated with the plan of exiting the deep economic crisis of the late 2000s while trying to reduce unemployment and poverty and commute to a low-carbon economy—all at the same time. Undoubtedly ambitious, it led to the sixth report, three years later, in which a better understanding of public expenditure and needed policy reforms would be developed.

The sixth cohesion report, published in 2014, emphasized further stimulation of investment for jobs and growth, promoted development and sound governance EU-wide with emphasis on cities. Unemployment, poverty, and exclusion were problems for all European regions and cities due to the continued and lingering aftereffects of the world crisis. However, economic disparities at the regional level had started to decline. Cohesion policy, connected with the Europe 2020 Strategy, led to the emerging inter-relating fields of smart, green, inclusive, sustainable growth. These ideas evolved from the foreseen impact of creating economic growth in conjunction with new jobs and improved effectiveness of good governance. The report accentuated the role of the European Structural and Investment Funds program by helping local areas grow. The funds supported investment in innovation, business, skills, and employment with emphasis on Europe 2020 targets. Investments using the funds focused on a low-carbon economy, innovation, and small-medium enterprises (SMEs). Moreover, quality employment, labor mobility, as well as social inclusion (i.e., in conjunction with digital networks, education, training, lifelong learning, and reform of public administration) played a significant role in the evolution from the fifth to sixth report [11, 12].

For the projected period of 2014–2020, some 88 programs were put forth to encourage integration among diffing policies, funds, and targets. The sixth cohesion report worked as a guide for the entirety of the venture capital enthused, profoundly impacting on national and regional budgets as well as limiting funding availability across all investment areas. Without cohesion policy, investments in MSs mostly affected by the economic crisis of 2008 would have fallen by an additional 50% [13]. In 2014, cohesion funding represented more than 60% of the investment budget in most eastern MSs. A strong inclination toward investment in SMEs—plus growth in RTD—raised innovation and skill levels to positively affect long-term structural obstacles. Cohesion policy, expected to bring a 2% GDP average growth and 1% higher level of employment in beneficiary MSs, was formulated with the idea of establishing functioning, accessible, and quality-oriented public services, to facilitate evidence-based policy making and deliver policy jointly with social partners and civil society [13]. These developments, in cooperation with regional investment, aimed at improving business competitiveness which forced heavy infrastructural advancements as an initiative to the seventh cohesion report.

The seventh cohesion report, published in 2017, is the latest edition of the report and last before 2020. It presented up-to-date results from the previous reports and additional clarity and objectivity on what has been achieved and what needs to be done in the post-2020 era. Average economic recovery seems to have taken root, GDP and employment have reached record levels and regional disparities shrinking. Unemployment rates remain above the pre-financial crisis level in a number of areas, while many SMEs struggle to "adapt to globalization, digitalization, green growth, and technology change" [14]. Concerns for public investment remain high, especially in MSs which suffered the most by the recent financial crisis, to the extent that cohesion funding remains a lifeline for many of them. The report also highlights the importance of continuing to modernize public institutions and implement necessary structural reforms to make them more efficient. "The report shows, in addition and without any ambiguity, that cohesion policy provided much needed help to MSs and regional and local authorities in the midst of the worst economic crisis thanks to its long-term, stable, and predictable investment" [14]. Key aspects of the report focalize on the impact of cohesion policy and measures to improve the effectiveness of programs introduced for the 2014–2020 period, including (1) ex-ante conditions, to stimulate structural reforms and to increase administrative capacity, (2) smart specialization strategies to identify local potential and prioritize investment in key sectors, and (3) a focus on results by program setting specific objectives and clear indicators of achievement. Targets for the 2014–2020 period include

- 14.5 million additional households with broadband access;
- 17 million additional people connected to wastewater facilities;
- 4600 km of renovated Trans-European Transport Network railway line;
- 6.8 million children with access to new or modernized schools;
- 7.4 million unemployed helped into work; and
- support for over 1million SMEs (i.e., to establish an additional 30,000 new research positions to bring new products to the market).

The establishment of the seventh cohesion report is the foundation of the latest results of this chapter and the premise information for whether the Europe 2020 targets will be met.

2.2   Implementation of Cohesion Policy: Europe 2020

To achieve the Europe 2020 targets MSs, in conjunction with regional areas, prepare structural reforms, looking for the best investment priorities to ensure micro and macrostability. The financial crisis in 2008 highly influenced the European economy compared with stability in previous periods [15]. The symptoms have been the declination in GDP, increment in unemployment, widespread poverty and social exclusion, rising general government debt and deficit, and regional disparities. As such, these problems have been a challenge for the EU's cohesion policy (i.e., specifically, its overall structural funding) for the period of 2014–2020 (Fig. 1).

Dating back to 2007–2013 period, cohesion policy effected economic growth and jobs, supported sustaining public expenditure (i.e., in vital areas like RTD) and SMEs, sustainable energy, human resources development and social inclusion, reforms in education, labor market, and public administration. This policy provided support for over 60,000 RTD projects, financing 50% of the total eligible costs by the end of 2012.

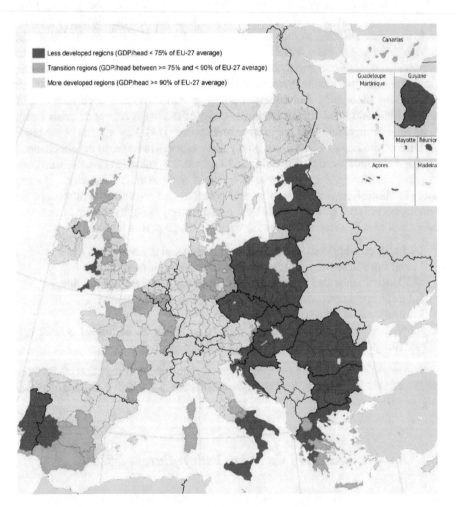

Fig. 1 Eligibility of structural funds from 2014 to 2020. Adapted from Eurostat [16]

These included 21,500 cooperation ventures between enterprises and research centers and 80,000 business start-ups. Vitally, it also provided funding to improve access to broadband Internet, supply of water, and drainage and connection to wastewater treatment facilities for about 5 million, 3.3 million, and 5.5 million people, respectively. Moreover, cohesion policy highly influenced the labor market, supporting 68 million participants in associated labor programs (i.e., 35 million women, 21 million younger adults, and 27 million low-level educated) and over 400,000 business start-ups for self-employment [11].

The EU budget for cohesion policy for the period of 2014–2020 is the current program being used to overcome disparities among MSs and regions striving to achieve the Europe 2020 Strategy targets. The targets are broken down into five overarching points and formulate the basis for the comparative results:

- to raise the employment rate of the population aged 20–64 from the current 69% to at least 75%;
- to achieve the target of investing 3% of GDP in RTD, in particular via the private sector;
- to reduce greenhouse gas emissions by at least 20% compared to 1990 levels or by 30% if the conditions are correct, increase the share of renewable energy in final energy consumption to 20%, and achieve a 20% increase in energy efficiency (i.e., -20, -20, -20 target);
- to reduce the share of early school leavers to 10% from the current 15% and increase the share of the population aged 30–34 having completed tertiary education from 31% to at least 40%; and
- to reduce the number of Europeans living below national poverty lines by 25%, lifting 20 million people out of poverty.

The Europe 2020 Strategy in terms of regional development is important for stimulating growth potential in regions otherwise under-supported. Every MS and region is encouraged to maximize the effects of investment through economic policy and sound regulatory, administrative, and institutional frameworks. Synergy of all these effects will bring better investment results and play a vital role in achieving strong cohesive unity.

An important challenge for the 2014–2020 period is for the improvement of institutional capacity and public administration that previously caused low government efficiency and insufficient absorption of money from the Structural and Cohesion Funds during the 2007–2013 cycle [17, 18]. New policy should overcome obstacles of economic development by moving away from low-level innovation, bettering labor skills as well as technical and transport infrastructure, and increasing institutional quality. This, in turn, can decrease growth, productivity, and standard of living. As such, from the previous seven-year cycle, cohesion policy has predominantly shifted in two directions: (1) from infrastructure investment to business support and innovation and (2) employment and social inclusion to overcoming barriers focused on less developed regions [5]. To better understand this transformation, an examination of financial sources, citing the two periods, is assessed.

2.3 Financial Sources for Cohesion Policy

The EU allocates funds for chosen targets in which MSs utilize at an individualistic level. For the two periods of 2007–2013 and 2014–2020, financial sources for cohesion policy are reviewed to better understand trends of convergence, regional competitiveness and employment (RCE), territorial cooperation, funded total support, and amount of support per capita.

In the 2007–2013 period, the largest amount of money spent on cohesion policy was in regions that covered convergence-oriented targets. Much of the funds were allocated on convergence and RCE compared to territorial cooperation targets— which closely align with the Lisbon Strategy as well as the Europe 2020 Strategy (Table 1). The main receivers of cohesion policy were MSs with less developed regions (i.e., GDP per capita < 75% of the MSs' average) and transition regions (i.e., GDP per capita between 75% and 90% of the MSs' average) [19]. Thus, support for cohesion policy is seen to be oriented toward Central and Eastern European (i.e., post-communist) countries and southern peripheries (i.e., Greece, Italy, Spain, and Portugal). Many parts of these MSs' so-called "peripheries" are rural areas, with relatively high unemployment, low income per capita, and non-competitive industry.

As such, the predominate rule for such funds is to support domestic resources that can cope with parallel changes of modernization and territorial equalization [20, 21]. The most important problem for these policies is to overcome territorial inequality. For the 2014–2020 period, much of the financial allocation, to date, has been allocated to regionally supporting lesser developed areas (Table 2). Cohesion funding has allotted additional resources to territorial cooperation and youth employment initiatives. During the initial phase of the 2014–2020 period, cohesion policy focalized on two targets: investment on growth and jobs and enhanced territorial cooperation. A large proportion of the budget was allocated in this manner for job creation. The largest budgetary allocation has been apportioned to Poland followed by Italy and Spain. These allotted funds support the Europe 2020 Strategy and illustrate important financial direction cohesion policy can stimulate for both the short- and long-term.

3 Achieving Europe 2020 Targets

Based on the observation of published Eurostat data, it is possible to consider which MSs will achieve the Europe 2020 targets. The premise in achieving these targets dates back to a ten-year strategy proposed by the European Commission on March 3, 2010, which closely followed the Lisbon Strategy from the decade early (i.e., 2000–2010). Results of MSs are grouped based on territorial and cultural closeness (i.e., Benelux = Belgium, Netherlands, and Luxembourg; Scandinavia = Denmark, Finland, and Sweden; Northern Western countries = Austria, France, Germany, Ireland, and the United Kingdom; Central and Eastern Europe = Czech Republic, Hungary, Poland, and Slovakia; Balkan countries = Bulgaria, Croatia, Romania, and

Table 1 Financial allocation in the EU's cohesion policy for the period of 2007–2013 (€ million)

Country	Convergence	Regional competitiveness and employment	Territorial cooperation	Total support	Amount of support per capita (€ million)
Austria	177	1027	257	1461	176
Belgium	638	1425	194	2258	213
Bulgaria	6674	–	179	6853	892
Cyprus	213	399	28	640	822
Czech Republic	25,883	419	389	26,692	2594
Denmark	–	510	103	613	112
Estonia	3404	–	52	3456	2574
Finland	–	1596	120	1716	325
France	3191	10,257	872	14,319	225
Germany	16,079	9409	851	26,340	320
Greece	19,575	635	210	20,420	1828
Hungary	22,890	2031	386	25,307	2514
Ireland	–	750	151	901	209
Italy	21,641	6325	846	28,812	487
Latvia	4531	–	90	4620	2025
Lithuania	6775	–	109	6855	2025
Luxembourg	–	50	15	65	136
Malta	840	–	15	855	2096
Netherlands	–	1660	247	1907	116
Poland	66,553	–	731	67,284	1765
Portugal	20,473	938	99	21,511	2029
Romania	19,213	–	455	19,668	912
Slovakia	10,912	–	227	11,588	2148
Slovenia	4101	–	104	4205	2092
Spain	26,180	8477	559	35,217	792
Sweden	–	1626	265	1891	207
United Kingdom	2912	6979	722	10,613	275

Adapted from Pastuszka [7]

Table 2 Financial allocation in the EU's cohesion policy for the period of 2014–2020 (€ million)

Country	Convergence	Regional competitiveness and employment	Territorial cooperation	Total support	Amount of support per capita (€ million)
Austria	–	978.3	257.3	–	1253.6
Belgium	638	1978.3	263.2	42.4	2283.9
Bulgaria	2278.3	5089.3	179.0	165.7	7588.4
Cyprus	269.5	421.8	32.8	11.6	735.6
Czech Republic	6258.9	15,370.7	339.7	13.6	21,982.9
Denmark	–	362.5	103.0	226.9	553.4
Estonia	1073.3	2461.2	55.4	–	3590.0
Finland	–	1304.4	120.0	161.3	1465.8
France	–	14,452.9	1089.3	310.2	15,852.5
Germany	–	18,269.5	851.0	965.4	19,234.9
Greece	3250.2	11,868.5	231.7	171.5	15,521.9
Hungary	6025.4	15,468.9	361.8	49.8	21,905.9
Ireland	–	951.6	168.8	68.1	1188.6
Italy	–	31,118.8	1136.7	567.5	32,823.0
Latvia	1349.4	3039.8	90.0	93.6	4511.8
Lithuania	2048.9	4628.7	113.8	31.8	6823.1
Luxembourg	–	39.6	20.2	–	59.7
Malta	217.7	490.2	17.0	–	725.0
Netherlands	–	1014.6	247.0	389.7	1404.3
Poland	23,208.0	53,406.0	700.5	252.4	77,567.0
Portugal	2861.7	18,320.0	122.4	160.8	21,465.0
Romania	6935.0	15,500.1	452.7	106.0	22,993.8
Slovakia	4168.3	9527.9	223.4	72.2	13,991.7
Slovenia	895.4	2107.3	62.9	9.2	3074.8
Spain	–	26,998.4	617.6	943.5	28,599.5
Sweden	–	1719.3	342.3	44.2	2105.8
United Kingdom	–	10,768.2	865.6	206.1	11,839.9

Adapted from Nitszke [18]

Slovenia; Baltic countries = Estonia, Latvia, and Lithuania; and Southern countries = Cyprus, Greece, Italy, Malta, Portugal, and Spain) for all target indicators. Tabular data is organized into three blocks (i.e., 2011–2013, 2014–2016, and 2017–2018 and target value).

Socioeconomic indicators are reviewed for five targets: rate of employment (i.e., for the age group 20–64) with the target set at 75%; percentage of expenditure on RTD with the target of 3%; reduction of tertiary education attainment (i.e., for the age group 30–34) with the target of below 40%; early leavers from education and training with the target of below 10%; and people at risk of poverty or social exclusion with the target of 20 million fewer people (Tables 3, 4, and 5).

The best socioeconomic results were obtained from MSs from Scandinavia followed by Benelux and Northern Western countries. Southern countries recorded the worst results. The source of this variation is reflective of economic policy, institutional structure, and specific societal norms in the given MS. The effectiveness of achieving the Europe 2020 targets over the entire period of 2011–2018 is also regionally compared using the average for each group of MSs. The leading MSs in terms of employment priorities were Scandinavia and the worst were Southern countries (Fig. 2). The assessment of expenditure on RTD confirmed a similar result with Scandinavia the only grouped MSs reaching the targeted level (Fig. 3). In terms of tertiary education attainment, for the age group 30–34, the highest positions were the Baltic countries, Benelux, and Scandinavia, while the worse were the Balkan countries (Fig. 4). For early leavers, most MSs—all-round—fulfilled this target with the best results coming from Central and Eastern Europe, Scandinavia, and Benelux. Underachievement with early leavers education and training was still prevalent in Southern countries (Fig. 5).

Risk of poverty or social exclusion was very low in Scandinavia and Benelux; however, throughout the Balkan countries, the recommended target of achieving 20 million fewer people from start to finish lagged behind and would require further cohesive maneuvering to better overall circumstances (Fig. 6).

Environmental indicators are illustrated for greenhouse gas emissions with the target of 20% (or 30% if the conditions are correct) lower than the 1990 level as well as share-level of renewable energy in gross final consumption with a target of 20% of energy from renewables (Tables 6, 7, and 8).

The best environmental results were found within the Baltic countries and Scandinavia, while Southern countries recorded the poorest results. Regionally, grouped MSs' greenhouse gas emission levels have overwhelmingly been met by the Baltic countries as well as the Balkan countries, Central and Eastern Europe, and Scandinavia—contrary, Northern Western countries and Benelux are approaching the target, while Southern countries are far from achieving it (Fig. 7).

An energy assessment on the percentage of shared renewable energy from gross final consumption for grouped MSs during the period of 2011–2017 Scandinavia, the Baltic countries, and the Balkan countries has met the target. All other grouped MSs fall short; however, Northern Western countries and Southern countries are approaching the target, while Benelux is far from achieving it (Fig. 8).

Table 3 Europe 2020 socioeconomic targets for MSs, 2011–2013

Grouped MSs (average)[a]	Rate of employment for the age group 20–64: target = 75%			Percentage of expenditure on RTD from GDP = target = 3%			Reduction of tertiary education attainment for the age group 30–34: target = below 40%			Early leavers education and training: target = below 10%			People at risk of poverty or social exclusion (%): target = 20 million fewer people		
	2011	2012	2013	2011	2012	2013	2011	2012	2013	2011	2012	2013	2011	2012	2013
A	71.5	71.9	71.6	1.8	1.8	1.8	43.9	45.2	46.1	9.2	9.6	8.8	17.8	18.3	18.6
Belgium	67.3	67.2	67.2	2.2	2.2	2.3	42.6	43.9	42.7	12.3	12.0	11.0	21.0	21.6	20.8
Netherlands	77.0	77.2	76.5	1.9	2.0	2.0	41.1	42.2	43.1	9.1	8.8	9.2	15.7	15.0	15.9
Luxembourg	70.1	71.4	71.1	1.4	1.2	1.2	48.2	49.6	52.5	6.2	8.1	6.1	16.8	18.4	19.0
B	76.3	76.3	76.2	3.3	3.2	3.2	44.7	45.6	45.6	8.7	8.5	8.1	18.0	17.5	17.5
Denmark	75.7	75.4	75.6	2.9	3.0	3.0	41.2	43.0	43.4	9.6	9.1	8.0	17.6	17.5	18.3
Finland	73.8	74.0	73.3	3.6	3.4	3.3	46.0	45.8	45.1	9.8	8.9	9.3	17.9	17.2	16.0
Sweden	79.4	79.4	79.8	3.2	3.3	3.2	46.8	47.9	48.3	6.6	7.5	7.1	18.5	17.7	18.3
C	71.6	71.9	72.5	2.2	2.2	2.2	38.7	40.1	40.9	11.5	10.6	9.5	22.1	22.3	22.4
Austria	75.2	75.6	75.5	2.7	2.8	2.8	23.8	26.3	27.3	8.3	7.6	7.3	19.2	18.5	18.8
France	69.3	69.4	69.9	2.2	2.2	2.2	43.3	43.5	44.1	11.9	11.5	9.7	19.3	19.1	18.1
Germany	76.3	76.7	77.1	2.8	2.9	2.9	30.7	32.0	33.1	11.7	10.6	9.9	19.9	19.6	20.3
Ireland	63.8	63.7	65.5	1.5	1.6	1.6	49.7	51.1	52.6	10.8	9.7	8.4	29.4	30.3	29.9
United Kingdom	73.6	74.2	74.9	1.7	1.6	1.6	45.8	47.1	47.6	15.0	13.6	12.4	22.7	24.1	24.8
D	65.3	65.9	66.4	1.0	1.2	1.3	27.9	29.6	31.5	6.7	7.0	7.3	23.6	24.0	23.7
Czech Republic	70.9	71.5	72.5	1.6	1.8	1.9	23.7	25.6	26.7	4.9	5.5	5.4	15.3	15.4	14.6
Hungary	60.7	62.1	63.2	1.2	1.3	1.4	28.1	29.9	31.9	11.2	11.5	11.8	31.5	33.5	34.8

(continued)

Table 3 (continued)

Grouped MSs (average)[a]	Rate of employment for the age group 20–64: target = 75%			Percentage of expenditure on RTD from GDP = target = 3%			Reduction of tertiary education attainment for the age group 30–34: target = below 40%			Early leavers education and training: target = below 10%			People at risk of poverty or social exclusion (%): target = 20 million fewer people		
	2011	2012	2013	2011	2012	2013	2011	2012	2013	2011	2012	2013	2011	2012	2013
Poland	64.5	64.7	64.9	0.8	0.9	0.9	36.5	39.1	40.5	5.6	5.7	5.6	27.2	26.7	25.8
Slovakia	65.0	65.1	65.0	0.7	0.8	0.8	23.2	23.7	26.9	5.1	5.3	6.4	20.6	20.5	19.8
E	62.8	62.6	62.9	1.1	1.1	1.1	27.5	27.9	29.5	9.4	9.6	9.6	35.5	36.2	36.0
Bulgaria	62.9	63.0	63.5	0.6	0.6	0.7	27.3	26.9	29.4	11.8	12.5	12.5	49.1	49.3	48.0
Croatia	57.0	55.4	57.2	0.8	0.8	0.8	24.5	23.7	25.6	4.1	4.2	4.5	32.6	32.6	29.9
Romania	62.8	63.8	63.9	0.5	0.5	0.4	20.4	21.8	22.8	17.5	17.4	17.3	40.9	43.2	41.9
Slovenia	68.4	68.3	67.2	2.4	2.6	2.6	37.9	39.2	40.1	4.2	4.4	3.9	19.3	19.6	20.4
F	67.9	69.6	70.9	1.3	1.3	1.1	40.6	41.8	45.2	9.9	9.1	8.6	32.1	30.7	29.8
Estonia	70.6	72.2	73.3	2.3	2.2	1.7	40.2	39.5	43.7	10.6	10.3	9.7	23.1	23.4	23.5
Latvia	66.3	68.1	69.7	0.7	0.7	0.6	35.9	37.2	40.7	11.6	10.6	9.8	40.1	36.2	35.1
Lithuania	66.9	68.5	69.9	0.9	0.9	1.0	45.7	48.6	51.3	7.4	6.5	6.3	33.1	32.5	30.8
G	64.4	62.5	61.5	0.9	0.9	0.9	31.3	32.8	33.9	19.1	17.8	16.6	26.2	27.9	28.4
Cyprus	73.4	70.2	67.2	0.5	0.4	0.5	46.2	49.9	47.8	11.3	11.4	9.1	24.6	27.1	27.8
Greece	59.6	55.0	52.9	0.7	0.7	0.8	29.1	312.0	34.9	12.9	11.3	10.1	31.0	34.6	35.7
Italy	61.2	61.0	59.8	1.2	1.3	1.3	20.3	21.7	22.4	18.2	17.6	17.0	28.1	29.9	28.5
Malta	61.6	63.1	64.8	0.7	0.9	0.9	23.4	24.9	26.0	22.7	21.1	20.8	22.1	23.1	24.0
Portugal	68.8	66.3	65.4	1.5	1.4	1.4	26.7	27.8	30.0	23.0	20.5	18.9	24.4	25.3	27.5
Spain	62.0	59.6	58.6	1.3	1.3	1.2	41.9	41.5	42.3	26.3	24.7	23.6	26.7	27.2	27.3

[a] A Benelux, B Scandinavia, C Northern Western countries, D Central and Eastern Europe, E Balkan countries, F Baltic countries, G Southern countries
Source Eurostat [16]

Table 4 Europe 2020 socioeconomic targets for MSs, 2014–2016

Grouped MSs (average)[a]	Rate of employment for the age group 20–64: target = 75%			Percentage of expenditure on RTD from GDP: target = 3%			Reduction of tertiary education attainment for the age group 30–34: target = below 40%			Early leavers education and training: target = below 10%			People at risk of poverty or social exclusion (%): target = 20 million fewer people		
	2014	2015	2016	2014	2015	2016	2014	2015	2016	2014	2015	2016	2014	2015	2016
A	71.6	71.5	71.8	1.9	1.9	1.9	47.1	47.1	48.6	8.2	9.2	7.4	18.9	18.7	19.1
Belgium	67.3	67.2	67.7	23.9	24.7	24.9	43.8	42.7	45.6	9.8	10.1	8.8	21.2	21.1	20.7
Netherlands	75.4	76.4	77.1	20.0	20.0	20.3	44.8	46.3	45.7	8.7	8.2	8.0	16.5	16.4	16.7
Luxembourg	72.1	70.9	70.7	12.6	12.7	12.4	52.7	52.3	54.6	6.1	9.3	5.5	19.0	18.5	19.8
B	76.5	76.6	77.3	3.1	3.0	2.9	46.7	47.8	48.3	8.0	8.0	7.5	17.8	17.7	17.2
Denmark	75.9	76.5	77.4	29.1	29.6	28.7	44.9	47.6	47.7	7.8	7.8	7.2	17.9	17.7	16.8
Finland	73.1	72.9	73.3	31.7	29.0	27.5	45.3	45.5	46.1	9.5	9.2	7.9	17.3	16.8	16.6
Sweden	80.5	80.5	81.2	31.5	32.7	32.5	49.9	50.2	51.0	6.7	7.0	7.4	18.2	18.6	18.3
C	73.1	73.7	74.5	2.3	2.2	2.2	43.0	43.2	43.5	9.4	8.8	8.8	22.0	21.1	20.5
Austria	74.2	74.3	74.8	30.7	30.5	30.9	40.0	38.7	40.1	7.0	7.3	6.9	19.2	18.3	18.0
France	69.3	69.5	70.0	22.3	22.7	22.5	43.7	45.0	43.6	11.9	9.0	9.2	18.5	17.7	18.2
Germany	77.7	78.0	78.6	28.7	29.2	29.4	31.4	32.3	33.2	9.5	10.1	10.3	20.6	20.0	19.7
Ireland	68.1	69.9	71.4	15.0	12.0	11.8	52.2	51.9	52.5	6.9	7.0	6.2	27.7	26.0	24.2
United Kingdom	76.2	76.8	77.5	16.7	16.7	16.9	47.7	47.9	48.2	11.8	10.8	11.2	24.1	23.5	22.2
D	68.1	69.8	70.6	1.3	1.4	1.2	32.8	34.1	35.5	7.2	7.5	7.9	22.4	21.0	19.9
Czech Republic	73.5	74.8	71.8	19.7	19.3	16.8	28.2	30.1	32.8	5.5	6.2	6.6	14.8	14.0	13.3
Hungary	66.7	68.9	71.5	13.5	13.6	12.1	34.1	34.3	33.0	11.4	11.6	12.4	31.8	28.2	26.3

(continued)

Table 4 (continued)

Grouped MSs (average)[a]	Rate of employment for the age group 20–64: target = 75%			Percentage of expenditure on RTD from GDP: target = 3%			Reduction of tertiary education attainment for the age group 30–34: target = below 40%			Early leavers education and training: target = below 10%			People at risk of poverty or social exclusion (%): target = 20 million fewer people		
	2014	2015	2016	2014	2015	2016	2014	2015	2016	2014	2015	2016	2014	2015	2016
Poland	66.5	67.8	69.3	9.4	10.0	9.7	42.1	43.4	44.6	5.4	5.3	5.2	24.7	23.4	21.9
Slovakia	65.9	67.7	69.8	8.8	11.8	7.9	26.9	28.4	31.5	6.7	6.9	7.4	18.4	18.4	18.1
E	64.4	65.7	66.4	10.8	1.1	10.3	32.2	32.9	33.2	9.5	10.1	10.0	32.5	31.7	31.4
Bulgaria	65.1	67.1	67.7	7.9	9.6	7.8	30.9	32.1	33.8	12.9	13.4	13.8	40.1	41.3	40.4
Croatia	59.2	60.6	61.4	7.8	8.4	8.5	32.1	30.8	29.3	2.8	2.8	2.8	29.3	29.1	27.9
Romania	65.7	66.0	66.3	3.8	4.9	4.8	25.0	25.6	25.6	18.1	19.1	18.5	40.3	37.4	38.8
Slovenia	67.7	69.1	70.1	23.7	22.0	20.0	41.0	43.4	44.2	4.4	5.0	4.9	20.4	19.2	18.4
F	72.3	74.1	75.0	1.1	1.0	0.9	45.5	48.1	48.9	8.8	9.2	8.6	28.7	28.1	27.7
Estonia	74.3	76.5	76.6	14.5	14.9	12.8	43.2	45.3	45.4	12.0	12.2	10.9	26.0	24.2	24.4
Latvia	70.7	72.5	73.2	6.9	6.3	4.4	39.9	41.3	42.8	8.5	9.9	10.0	32.7	30.9	28.5
Lithuania	71.8	73.3	75.2	10.3	10.4	8.5	53.3	57.6	58.7	5.9	5.5	4.8	27.3	29.3	30.1
G	62.4	63.7	65.1	1.0	1.0	0.9	35.6	36.8	37.8	15.1	13.5	13.4	28.7	28.5	27.7
Cyprus	67.6	67.9	68.7	5.1	4.8	5.0	52.5	54.5	53.4	6.8	5.2	7.6	27.4	28.9	27.7
Greece	53.3	54.9	56.2	8.3	9.7	10.1	37.2	40.4	42.7	9.0	7.9	6.2	36.0	35.7	35.6
Italy	59.9	60.5	61.6	13.4	13.4	12.9	23.9	25.3	26.2	15.0	14.7	13.8	28.3	28.7	30.0
Malta	66.4	67.8	69.6	7.2	7.7	6.1	26.5	27.8	29.9	20.3	19.8	19.7	23.8	22.4	20.1
Portugal	67.6	69.1	70.6	12.9	12.4	12.7	31.3	31.9	34.6	17.4	13.7	14.0	27.5	26.6	25.1
Spain	59.9	62.0	63.9	12.4	12.2	11.9	42.3	40.9	40.1	21.9	20.0	19.0	29.2	28.6	27.9

[a] A Benelux, B Scandinavia, C Northern Western countries, D Central and Eastern Europe, E Balkan countries, F Baltic countries, G Southern countries
Source Eurostat [16]

Table 5 Europe 2020 socioeconomic targets for MSs, 2017–2018 and target value

Grouped MSs (average)[a]	Rate of employment for the age group 20–64: target = 75%			Percentage of expenditure on R&D from GDP: target = 3%			Reduction of tertiary education attainment for the age group 30–34: target = below 40%			Early leavers education and training: target = below 10%			People at risk of poverty or social exclusion (%): target = 20 million fewer people		
	2017	2018	Target	2017	2018	Target	2017	2018	Target	2017	2018	Target	2017	2018	Target
A	72.7	73.7	75.0	1.9	–	3.0	48.8	51.1	40.0	7.8	7.4	10.0	19.6	18.3	20.0
Belgium	68.5	69.7	73.2	2.6	–	3.0	45.9	47.6	47.0	8.9	8.6	9.5	20.3	19.8	20.0
Netherlands	78.0	79.2	80.0	1.9	–	2.5	47.9	49.4	40.0	7.1	7.3	8.0	17.0	16.8	20.0
Luxembourg	71.5	72.1	73.0	1.3	–	2.3	52.7	56.2	66.0	7.3	6.3	10.0	21.5	–	20.0
B	77.6	79.0	75.0	3.0	–	3.0	48.2	47.8	40.0	8.2	9.3	10.0	16.9	17.4	20.0
Denmark	76.9	78.2	80.0	3.0	–	3.0	48.8	49.1	40.0	8.8	10.2	10.0	17.2	17.6	20.0
Finland	74.2	76.3	78.0	2.7	–	4.0	44.6	44.2	42.0	8.2	8.3	8.0	15.7	16.5	20.0
Sweden	81.8	82.6	80.0	3.4	–	4.0	51.3	52.0	45.0	7.7	9.3	7.0	17.7	18.0	20.0
C	75.3	76.0	75.0	2.2	–	3.0	44.4	45.4	40.0	8.4	8.4	10.0	19.8	–	20.0
Austria	75.4	76.2	77.0	3.2	–	3.76	40.8	40.7	38.0	7.4	7.3	9.5	18.1	17.5	20.0
France	70.6	71.3	75.0	2.2	–	3.0	44.3	46.2	50.0	8.9	8.9	9.5	17.1	–	20.0
Germany	79.2	79.9	77.0	3.0	–	3.0	34.0	34.9	42.0	10.1	10.3	7.0	19.0	–	20.0
Ireland	73.0	74.1	69.0	1.0	–	2.0	54.5	56.3	60.0	5.0	5.0	8.0	22.7	–	20.0
United Kingdom	78.2	78.7	80.0	1.7	–	4.0	48.3	48.8	45.0	10.6	10.7	11.2	22.0	–	20.0
D	73.4	74.7	75.0	1.2	–	3.0	36.6	37.7	40.0	8.3	8.0	10.0	25.8	19.5	20.0
Czech Republic	78.5	79.9	75.0	1.8	–	1.0	34.2	33.7	32.0	6.7	6.2	5.5	12.2	12.2	20.0
Hungary	73.3	74.4	75.0	1.3	–	1.8	32.1	33.7	34.0	12.5	12.5	10.0	25.6	19.6	20.0

(continued)

Table 5 (continued)

Grouped MSs (average)[a]	Rate of employment for the age group 20–64: target = 75%			Percentage of expenditure on R&D from GDP: target = 3%			Reduction of tertiary education attainment for the age group 30–34: target = below 40%			Early leavers education and training: target = below 10%			People at risk of poverty or social exclusion (%): target = 20 million fewer people		
	2017	2018	Target	2017	2018	Target	2017	2018	Target	2017	2018	Target	2017	2018	Target
Poland	70.9	72.2	71.0	1.0	–	1.7	45.7	45.7	45.0	5.0	4.8	4.5	19.5	18.9	20.0
Slovakia	71.1	72.4	72.0	0.9	–	1.2	34.3	37.7	40.0	9.3	8.6	6.0	16.3	–	20.0
E	69.3	70.7	75.0	1.0	–	10.3	32.2	32.9	33.2	9.5	9.1	10.0	29.5	26.6	20.0
Bulgaria	71.3	72.4	76.0	0.7	–	1.5	32.8	33.7	40.0	12.7	12.7	11.0	38.9	32.8	20.0
Croatia	63.6	65.2	62.9	0.9	–	1.4	28.7	34.1	35.0	3.1	3.3	4.0	26.4	24.9	20.0
Romania	68.8	69.9	70.0	0.5	–	2.0	26.3	24.5	26.7	18.1	16.4	11.3	35.7	32.5	20.0
Slovenia	73.4	75.4	75.0	1.9	–	3.0	46.4	42.7	40.0	4.3	4.2	5.0	17.1	16.2	20.0
F	76.5	78.0	75.0	0.9	–	0.9	50.0	49.2	48.9	8.3	8.1	10.0	27.1	26.4	20.0
Estonia	78.7	79.5	76.0	1.3	–	3.0	48.4	47.2	40.0	10.8	11.3	9.5	23.4	24.4	20.0
Latvia	74.8	76.8	73.0	0.5	–	1.5	43.8	42.7	34.0	8.6	8.3	10.0	28.2	28.4	20.0
Lithuania	76.0	77.8	72.8	0.9	–	1.9	58.0	57.6	48.7	5.4	4.6	9.0	29.6	–	20.0
G	67.1	68.9	75.0	1.0	–	0.9	39.1	39.9	40.0	12.8	12.4	10.0	26.3	24.6	20.0
Cyprus	70.8	73.9	75.0	0.6	–	0.5	55.9	57.1	46.0	8.5	7.8	10.0	25.2	–	20.0
Greece	57.8	59.5	70.0	1.1	–	1.2	43.7	44.3	32.0	6.0	4.7	10.0	34.8	31.8	20.0
Italy	62.3	63.0	67.0	1.3	–	1.5	26.9	27.8	26.0	14.0	14.5	16.0	28.9	–	20.0
Malta	73.0	75.0	70.0	0.5	–	2.0	33.5	34.2	33.0	17.7	17.5	10.0	19.3	19.0	20.0
Portugal	73.4	75.4	75.0	1.3	–	2.7	33.5	33.5	40.0	12.6	11.8	10.0	23.3	21.6	20.0
Spain	65.5	67.0	74.0	1.2	–	2.0	41.2	42.4	44.0	18.3	17.9	15.0	26.6	26.1	20.0

[a] A Benelux, B Scandinavia, C Northern Western countries, D = Central and Eastern Europe, E = Balkan countries, F Baltic countries, G Southern countries
Source Eurostat [16]

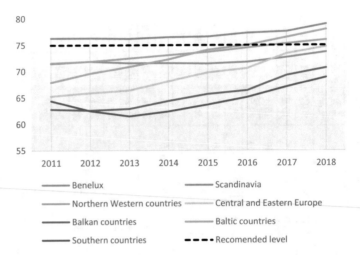

Fig. 2 Rate of employment for the age group 20–64 for grouped MSs. Adapted from Eurostat [16]

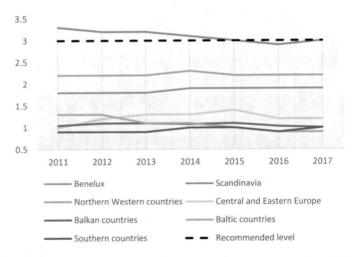

Fig. 3 Expenditure on RTD as a percentage of GDP for grouped MSs from 2011 to 2107. Adapted from Eurostat [16]; *Note* Data is not available for 2018

It is worth noting that regional diversity, within the context of the Europe 2020 indicators, is attributed to the condition of the economy, degree of innovation, and technical development within each MS. Regions that recorded low results are encouraged to reform economic and institutional frameworks to reinforce their position and develop competition, including better economic development (i.e., more employment with a reciprocal policy for less risk of poverty and social exclusion), effective institutional deregulation, avoidance of bureaucracy, institutional support for innovation, decreased risk of commercialization of new technology and business ideas,

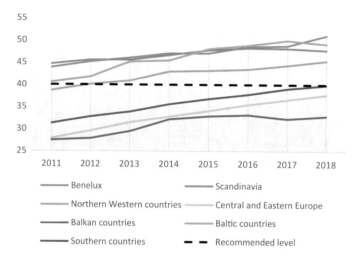

Fig. 4 Tertiary education attainment for the age group 30–34 for grouped MSs from 2011 to 2018. Adapted from Eurostat [16]

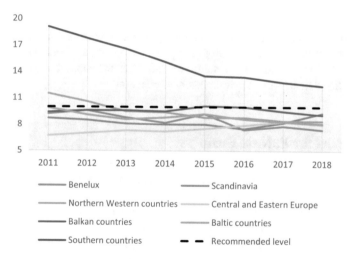

Fig. 5 Early leavers education and training for grouped MSs from 2011–2018. Adapted from Eurostat [16]

and elasticity of the education system [22–24]. Reforms that mimic these practices would offer additional opportunities toward permanent learning for a changing global economy and order [24, 25].

Issues concerning climate and energy targets are mostly tech-driven [26–29]. At present, most technological advancements require expensive fixes and increased

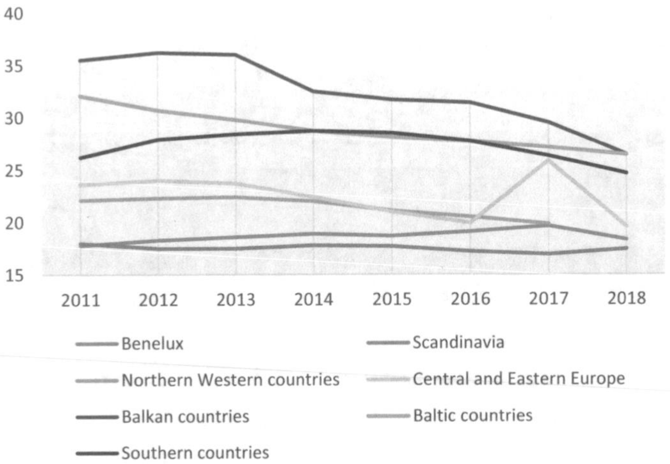

Fig. 6 Number of people at risk of poverty or social exclusion (i.e., a target of 20 million fewer people from start to finish) for grouped MSs from 2011 to 2018. Adapted from Eurostat [16]

industry costs which traditionally correlate with an increase in unemployment [30–33]. Similarly, raising capital for eco-friendly technology is another challenge [26, 27]. In terms of the Europe 2020 targets, private sector entities should be encouraged to invest in climate and energy solutions and look for support mechanisms and opportunities that can offset private business concerns.

4 Support for Private Business

It is a challenging task to compare MSs solely on their rank and capacity of doing business; however, an analysis of private sector entities indicates some of the Europe 2020 targets parallel similar means. Support for entrepreneurial development can be one potential sign. As such, ranked data integrating positive economic incentives shows corresponding socioeconomic and environmentally friendly (i.e., eco-oriented) business overlaps with cohesion-oriented targets. According to the World Bank's [34] database on doing business, a number of key factors should be considered when supporting private business and growth, they include starting a business, dealing with construction permits, getting electricity, registering property, getting credit, protecting minority investors, paying taxes, trading across borders, enforcing contracts, and resolving insolvency. Utilizing these factors, world ranking results are presented for the EU in Table 9.

Within the EU, Scandinavia and the Baltic countries are among the best ranked in terms of creating a favorable environment for doing business. Other Western countries closely follow with the United Kingdom notably being the second most favorable MS after Denmark. Conditions for private business (i.e., ease of doing business)

Table 6 Europe 2020 environmental targets for MSs, 2011–2013

Grouped MSs (average)[a]	Greenhouse gas emissions (i.e., base year = 1990): target = 20% (or 30% if the conditions are correct) lower than 1990			Share of renewable energy in gross final consumption: target = 20% of energy from renewable		
	2011	2012	2013	2011	2012	2013
A	92.7	90.2	88.9	4.6	5.0	5.3
Belgium	84.5	82.4	82.5	6.3	7.2	7.5
Netherlands	93.0	90.6	90.7	4.5	4.7	4.8
Luxembourg	100.5	97.7	93.6	2.9	3.1	3.5
B	88.8	82.5	83.0	35.0	37.1	38.7
Denmark	84.0	77.5	79.9	23.5	25.7	27.4
Finland	96.5	88.8	89.9	32.8	34.4	36.7
Sweden	85.9	81.2	79.2	48.8	51.1	52.0
C	89.9	89.8	89.9	12.8	13.7	14.5
Austria	106.4	103.0	103.3	30.6	31.5	32.4
France	90.3	90.3	90.2	11.1	13.4	14.1
Germany	74.6	75.2	76.6	11.4	12.1	12.4
Ireland	104.6	105.2	105.4	6.5	7.1	7.7
United Kingdom	73.6	75.5	73.8	4.2	4.6	5.7
D	71.6	68.9	67.2	11.4	12.4	12.9
Czech Republic	69.7	67.6	65.0	10.9	12.8	13.8
Hungary	68.4	64.3	61.3	14.0	15.5	16.2
Poland	86.9	85.4	84.7	10.3	10.9	11.4
Slovakia	61.3	58.3	57.7	10.3	10.4	10.1
E	76.8	73.1	69.0	20.3	21.6	23.3
Bulgaria	63.4	58.5	53.5	14.3	16.0	19.0
Croatia	86.2	80.7	76.9	25.4	26.8	28.0
Romania	51.8	50.6	46.8	21.4	22.8	23.9
Slovenia	105.7	102.6	98.9	20.3	20.8	22.4
F	47.1	46.0	46.5	26.3	27.6	28.5
Estonia	52.5	49.8	54.2	25.5	25.8	25.6
Latvia	44.7	44.1	43.8	33.5	35.7	37.1
Lithuania	44.2	44.1	41.5	19.9	21.4	22.7
G	125.7	123.3	114.6	11.6	12.9	14.1
Cyprus	158.0	149.2	137.2	6.0	6.8	8.1
Greece	111.8	108.4	99.4	10.9	13.5	15.0
Italy	95.8	91.9	86.1	12.9	15.4	16.7

(continued)

Table 6 (continued)

Grouped MSs (average)[a]	Greenhouse gas emissions (i.e., base year = 1990): target = 20% (or 30% if the conditions are correct) lower than 1990			Share of renewable energy in gross final consumption: target = 20% of energy from renewable		
	2011	2012	2013	2011	2012	2013
Malta	145.8	153.1	139.4	1.9	2.8	3.7
Portugal	116.7	113.7	110.8	24.6	24.6	25.7
Spain	126.1	123.8	114.6	13.2	14.3	15.3

[a]*A* Benelux, *B* Scandinavia, *C* Northern Western countries, *D* Central and Eastern Europe, *E* Balkan countries, *F* Baltic countries, *G* Southern countries
Source Eurostat [16]

are a key factor in creating socioeconomic cohesion. In this manner, MSs should look for endogenous possibilities to develop policy which favor private business, industry innovation, and RTD. Some additional approaches include a low, simple, clear tax system; ease for economic activity; and eliminating or low corruption levels. Moreover, the contribution of simple economic laws, availability of high-quality public goods and services, an open and competitive market economy, and stable monetary policy are also paramount.

5   Discussion

A number of published papers have investigated the effectiveness of cohesion policy. Boldrin et al. [35] reported funds which support less developed MSs and regions have mostly a social impact versus developmental. Structural policy is restricted to redistribution targets and statistically does not confirm permanent accelerative growth. Sapir et al. [36] also critically argued that cohesion policy should always integrate with the fact that the EU's economy is less than the USA and should carefully take into account post-communist MSs' potential as well as China when determining indicators and targets. For example, high levels of employment can significantly influence economic growth and cause high labor productivity. On other hand, several reasons for slow stagnant growth include overregulation of business activities, high taxes, and high cost of living. Moreover, low-level innovation due to limited RTD confirms convergence requisites between MSs and divergence problems at the regional scale (i.e., especially within large cities). Recommendations would be to direct funds toward innovation and growth as well as develop rural policy while traditional, structural policy be lessened.

A lack of institutional frameworks can act as an obstacle in effectively absorbing EU structural support. Traditionally, the EU has not encouraged development possibilities for less developed MSs when underdeveloped institutional frameworks exist. As such, the European Structural and Investment Funds program provisions

Table 7 Europe 2020 environmental targets for MSs, 2014–2016

Grouped MSs (average)[a]	Greenhouse gas emissions (i.e., base year = 1990): target = 20% (or 30% if the conditions are correct) lower than 1990			Share of renewable energy in gross final consumption: target = 20% of energy from renewable		
	2014	2015	2016	2014	2015	2016
A	85.7	87.0	86.9	6.0	6.2	6.7
Belgium	78.8	81.5	81.5	8.0	7.9	8.7
Netherlands	87.4	91.3	91.6	5.5	5.8	6.0
Luxembourg	90.8	88.3	87.5	4.5	5.0	5.4
B	78.5	75.7	78.0	40.3	41.3	41.6
Denmark	74.4	70.9	73.9	29.6	31.0	32.2
Finland	84.1	79.3	84.0	38.7	39.2	38.7
Sweden	77.1	76.8	76.1	52.5	53.8	53.8
C	86.2	87.5	87.9	15.4	16.0	16.6
Austria	98.6	101.8	103.1	33.0	32.8	33.5
France	84.8	85.7	85.6	14.7	15.1	16.0
Germany	73.4	73.7	74.0	13.8	14.6	14.8
Ireland	105.3	109.5	113.4	8.7	9.2	9.5
United Kingdom	68.8	66.7	63.6	7.0	8.5	9.3
D	65.7	67.0	68.0	13.2	13.5	13.1
Czech Republic	64.1	64.6	65.6	15.0	15.0	14.9
Hungary	62.0	65.3	65.8	14.6	14.4	14.2
Poland	81.9	82.7	85.0	11.5	11.7	11.3
Slovakia	54.8	55.4	55.6	11.7	12.9	12.0
E	66.8	68.3	68.7	23.0	23.5	23.3
Bulgaria	56.4	59.5	57.0	18.0	18.2	18.8
Croatia	74.3	75.8	76.2	27.8	29.0	28.3
Romania	46.8	47.2	45.8	24.8	24.8	25.0
Slovenia	89.6	90.7	95.2	21.5	21.9	21.3
F	45.7	43.5	44.8	29.4	30.7	30.5
Estonia	52.3	44.7	48.6	26.3	28.6	28.8
Latvia	43.4	43.7	43.8	38.7	37.6	37.2
Lithuania	41.5	42.1	42.0	23.6	25.8	25.6
G	115.1	111.9	109.7	14.8	15.2	15.6
Cyprus	143.3	143.8	152.9	8.9	9.4	9.3
Greece	96.5	93.0	89.7	15.3	15.4	15.2
Italy	83.1	84.7	83.8	17.1	17.5	17.4

(continued)

Table 7 (continued)

Grouped MSs (average)[a]	Greenhouse gas emissions (i.e., base year = 1990): target = 20% (or 30% if the conditions are correct) lower than 1990			Share of renewable energy in gross final consumption: target = 20% of energy from renewable		
	2014	2015	2016	2014	2015	2016
Malta	140.9	112.0	99.4	4.7	5.0	6.0
Portugal	111.0	118.3	115.8	27.0	28.0	28.5
Spain	115.6	119.7	116.4	16.1	16.2	17.3

[a]*A* Benelux, *B* Scandinavia, *C* Northern Western countries, *D* Central and Eastern Europe, *E* Balkan countries, *F* Baltic countries, *G* Southern countries
Source Eurostat [16]

first the implementation and maturation of institutional structures for a well-framed, top-down makeup before funds are considered [37]. Funds should finance the creation of institutional frameworks which positively can focus on economic growth and development. This situation, thereby, is meant to waste less financial resources regardless of socioeconomic level [38]. Cohesion policy needs to effectively allocate funds on the basis of a stable framing and target priority. As such, it is possible to improve policy via administrative reform, improved implementation, and reliability assessment of framework results [9, 39–42]. In spite of this, some factors between cohesion policy and the Europe 2020 Strategy need to be stringently connected, since lack of synergy would be an obstacle for establishing priorities [37, 43, 44].

A number of examples, in terms of intervention at regional and local levels, connect policy with infrastructure—particularly transport infrastructure. Several of the less developed MSs utilize financial support in building transport infrastructure. In theory, infrastructure investment brings immediate regional and national economic return [45–47]. Differing analyses, however, show this correlation not as concrete [45, 48]. Investment in transport infrastructure should be second to education. Transport infrastructure interconnects cities making urban life more viable as well as opportunistic. Critics however have implied excessive infrastructure investment can bring problems like overcapitalization and fragility via over-exploitation of resources [49–52]. Another concern is infrastructure maintenance and associated costs [53, 54]. To align toward a best practices approach, arguments that change and reorient cohesion policy, the Europe 2020 Strategy, and individual MS's economic policy should underscore the importance of both public and private sector entities to enhance competitiveness and innovation. The EU's multi-governmental hierarchy for creating development strategies—and associated policy—should be clear, achievable, and not contradictory (i.e., between development program and intended target) [2, 46, 52].

Table 8 Europe 2020 environmental targets for MSs, 2017 and target value

Grouped MSs (average)[a]	Greenhouse gas emissions (i.e., base year = 1990): target = 20% (or 30% if the conditions are correct) lower than 1990			Share of renewable energy in gross final consumption: target = 20% of energy from renewable		
	2017	2018	Target	2017	2018	Target
A	87.1	–	80.0	7.3	–	20.0
Belgium	79.7	–	80.0	9.0	–	20.0
Netherlands	90.9	–	80.0	6.6	–	20.0
Luxembourg	90.8	–	80.0	6.4	–	20.0
B	75.4	–	80.0	43.7	–	20.0
Denmark	70.5	–	80.0	35.7	–	20.0
Finland	79.5	–	80.0	41.0	–	20.0
Sweden	76.3	–	80.0	54.5	–	20.0
C	88.4	–	80.0	17.0	–	20.0
Austria	106.2	–	80.0	32.5	–	20.0
France	86.6	–	80.0	16.3	–	20.0
Germany	74.1	–	80.0	15.4	–	20.0
Ireland	112.9	–	80.0	10.6	–	20.0
United Kingdom	62.4	–	80.0	10.2	–	20.0
D	70.1	–	80.0	12.6	–	20.0
Czech Republic	65.3	–	80.0	14.8	–	20.0
Hungary	68.5	–	80.0	13.3	–	20.0
Poland	87.6	–	80.0	10.9	–	20.0
Slovakia	59.2	–	80.0	11.5	–	20.0
E	69.8	–	80.0	23.0	–	20.0
Bulgaria	60.5	–	80.0	18.7	–	20.0
Croatia	78.7	–	80.0	27.3	–	20.0
Romania	46.1	–	80.0	24.5	–	20.0
Slovenia	93.8	–	80.0	21.5	–	20.0
F	46.3	–	80.0	31.3	–	20.0
Estonia	51.9	–	80.0	29.2	–	20.0
Latvia	44.3	–	80.0	39.0	–	20.0
Lithuania	42.7	–	80.0	25.8	–	20.0
G	115.0	–	80.0	16.8	–	20.0
Cyprus	155.7	–	80.0	9.8	–	20.0
Greece	93.6	–	80.0	16.9	–	20.0
Italy	84.1	–	80.0	18.3	–	20.0

(continued)

Table 8 (continued)

Grouped MSs (average)[a]	Greenhouse gas emissions (i.e., base year = 1990): target = 20% (or 30% if the conditions are correct) lower than 1990			Share of renewable energy in gross final consumption: target = 20% of energy from renewable		
	2017	2018	Target	2017	2018	Target
Malta	112.2	–	80.0	7.1	–	20.0
Portugal	122.8	–	80.0	28.1	–	20.0
Spain	121.8	–	80.0	17.5	–	20.0

[a]A Benelux, B Scandinavia, C Northern Western countries, D Central and Eastern Europe, E Balkan countries, F Baltic countries, G Southern countries
Source Eurostat [16]
Note Data is not available for 2018

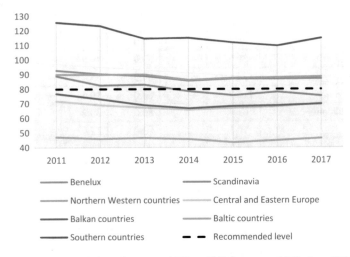

Fig. 7 Greenhouse gas emissions (base year 1990 = 100) for grouped MSs from 2011 to 2017. Adapted from Eurostat [16]; *Note* Data is not available for 2018

6 Conclusion

Cohesion policy is an important tool for harmonized development. It envisions decreased disparity and advancement among MSs, especially at the regional level. The results from the three cohesion reports (i.e., the fifth, sixth, and seventh) underline the initial importance of the Lisbon Strategy as a precursor to the Europe 2020 Strategy. These strategies should be prioritized for improved cohesion policy. For Europe 2020, best results were obtained by Scandinavia, Benelux, and Northern Western countries, while Southern countries recorded—overall—low results. It is noted, the creation of sound conditions for business will be critical for balanced

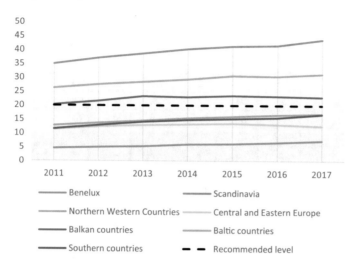

Fig. 8 Percentage share of renewable energy from gross final consumption for grouped MSs from 2011 to 2017. Adapted from Eurostat [16]; *Note* Data is not available for 2018

EU-wide outputs. Generally, there are mixed results on how European funds can be more effectively allocated. The fundamental concern would be to assess region- and MS-specific problems and allocate best fit requirements tailored to each. To date, improvements should safeguard effective, proper institutional frameworks; implemented transport infrastructure; systemically applied education; and robust, amplified innovation policy.

The Europe 2020 targets show a valuable backdrop into structuring proper cohesion policy as well as the potential for failed policy (i.e., both at the region- and MS-level). For example, in 2016, it can be argued that failed cohesion policy, the decade prior, led to the leavers winning the Brexit vote in the United Kingdom sparking similar EU-leave movements throughout the Union [55]. Also, the Visegrád Group, made up of the Czech Republic, Hungary, Poland, and Slovakia, opposes the majority of the MSs on migration policy with some reinstalling border fencing (i.e., to control migrant flow) [56–59]. As such, when considering these two examples, EU indicators and target values must contextualize present-day convergence and future needs as critical to effective cohesion policy and cohesiveness for a standing Union. At the European level, cohesion policy can be seen as challenging and strategically beneficial, while to the rest of the world, it is observed and assessed on its reciprocity.

Table 9 Doing business world ranking for the EU from 2011 to 2018

Grouped MSs	2011	2012	2013	2014	2015	2016	2017	2018
Benelux								
Belgium	27	28	42	36	42	43	42	52
Netherlands	29	31	27	28	27	28	28	32
Luxembourg	44	50	59	60	59	61	59	63
Scandinavia								
Denmark	5	5	4	5	4	3	3	3
Finland	14	11	9	12	9	10	13	13
Sweden	9	14	11	14	11	8	9	10
Other Western countries								
Austria	11	15	10	30	21	21	15	14
France	26	29	31	38	31	27	29	31
Germany	19	19	14	21	14	15	17	20
Ireland	8	10	13	15	13	17	18	17
United Kingdom	6	7	8	10	8	6	7	7
Central and Eastern Europe								
Czech Republic	70	64	44	75	44	36	27	30
Hungary	46	51	54	54	54	42	41	48
Poland	59	62	32	45	32	25	24	27
Slovakia	43	48	37	49	37	29	33	39
Balkan countries								
Bulgaria	57	59	38	58	38	38	39	50
Croatia	79	80	65	89	65	40	43	51
Romania	65	72	48	73	48	37	36	45
Slovenia	37	37	51	33	51	29	37	30
Baltic countries								
Estonia	18	24	17	22	17	16	12	12
Latvia	31	21	23	24	23	22	14	19
Lithuania	25	27	24	17	24	20	21	16
Southern countries								
Cyprus	49	40	64	39	64	47	53	45
Greece	101	100	61	72	61	60	61	67
Italy	83	87	56	65	56	45	50	46
Malta	–	–	–	103	94	80	76	84
Portugal	30	30	25	31	25	23	25	29
Spain	45	44	33	52	33	33	32	28

Adapted from the World Bank [34]

References

1. Molle W (2007) European cohesion policy. Routledge, London
2. European Commission (2017) White paper on the future of Europe and the way forward. European Commission, Brussels
3. Warlouzet L (2019) The EEC/EU as an evolving compromise between French Dirigism and German Ordoliberalism (1957–1995). J Common Mark Stud 57:77–93. https://doi.org/10.1111/jcms.12817
4. European Commission (2019) European neighbourhood policy and enlargement negotiations. In: European Commission https://ec.europa.eu/neighbourhood-enlargement/node_en. Accessed 28 Jul 2019
5. Bachtler J, Berkowitz P, Hardy S (2016) EU cohesion policy: reassessing performance and direction. Routledge, London
6. European Commission (2008) Consolidated version of the treaty on the functioning of the European Union: EUR-Lex—12008E174—EN—EUR-Lex. In: European Commission Law. https://eur-lex.europa.eu/legal-content/EN/TXT/?uri=CELEX%3A12008E174. Accessed 31 Jul 2019
7. Pastuszka S (2012) Polityka regionalna Unii Europejskiej. Difin, Warsaw
8. Commission European (2010) Fifth report on economic, social and territorial cohesion: investing in Europe's future. European Commission, Brussels
9. Kukuła AJ (2015) Cohesion policy and development of the European Union's regions in the perspective of 2020. Wydawnictwo KUL, Lublin
10. Rodriguez R, Warmerdam J, Triomphe CE et al (2010) The lisbon strategy 2000–2010: an analysis and evaluation of the methods used and results achieved. European Parliament, Brussels
11. Dijkstra L (2015) Investment for jobs and growth: publications office of the EU. European Commission, Brussels
12. Koczor M, Tokarski P (2011) From Lisbon to Europe 2020. Lisbon strategy implementation in 2010: assessments and prospects. The Polish Institute of International Affairs, Warsaw
13. European Commission (2014) Sixth report on economic, social and territorial cohesion. European Commission, Brussels
14. European Commission (2017) Seventh report on economic, social and territorial cohesion. European Commission, Brussels
15. Ágh A (2011) European Union at the crossroads: the European perspectives after the global crisis. Budapest College of Communication, Business and Arts, Budapest
16. Eurostat (2019) Eurostat database. In: European Commission. https://ec.europa.eu/eurostat/data/database. Accessed 31 Jul 2019
17. Marlier E, Natali D (2010) Europe 2020: towards a more social EU?. P.I.E, Peter Lang, Brussels
18. Nitszke A (2015) Polityki europejskie, perspektywa finansowa UE 2014-2020. Ministerstwo Rozwoju, Warsaw
19. Mohl P (2016) Empirical evidence on the macroeconomic effects of EU cohesion policy. Springer Fachmedien Wiesbaden, Wiesbaden
20. Kovács IP (2011) The future of European cohesion policy: the paradoxes in domestic governance of CEE countries. European Union at the crossroads: the European perspectives after the global crisis. Kossuth Kiadó Zt, Frankfurt, pp 213–233
21. Heinelt H, Petzold W (2018) The structural funds and EU cohesion policy. In: Heinelt H, Münch S (eds) Handbook of European Policies: interpretive approaches to the EU. Edward Elgar Publishing Limited, Cheltenham, pp 134–155
22. Erixon F (2010) The Europe 2020 strategy: time for Europe to think again. Eur View 9:29–37. https://doi.org/10.1007/s12290-010-0120-8
23. Brauers WKM, Baležentis A, Baležentis T (2012) European Union member states preparing for Europe 2020. An application of the Multimoora method. Technol Econ Dev Econ 18:567–587. https://doi.org/10.3846/20294913.2012.734692
24. Abbott A, Schiermeier Q (2019) How European scientists will spend €100 billion. Nat 2019:5697757

25. Erixon F (2009) SMEs in Europe: taking stock and looking forward. Eur View 8:293–300. https://doi.org/10.1007/s12290-009-0093-7
26. Kumar V, Kumar U (2017) Introduction: technology, innovation and sustainable development. Transnatl Corp Rev 9:243–247. https://doi.org/10.1080/19186444.2017.1408553
27. Mazzanti M (2018) Eco-innovation and sustainability: dynamic trends, geography and policies. J Environ Plan Manag 61:1851–1860. https://doi.org/10.1080/09640568.2018.1486290
28. Pearce J, Albritton S, Grant G et al (2012) A new model for enabling innovation in appropriate technology for sustainable development. Sustain Sci Pract Policy 8:42–53. https://doi.org/10.1080/15487733.2012.11908095
29. Geng JB, Ji Q (2016) Technological innovation and renewable energy development: evidence based on patent counts. Int J Glob Environ Issues 15:217. https://doi.org/10.1504/IJGENVI.2016.076945
30. Cascio WF, Montealegre R (2016) How technology is changing work and organizations. Annu Rev Organ Psychol Organ Behav 3:349–375. https://doi.org/10.1146/annurev-orgpsych-041015-062352
31. Parry E, Battista V (2019) The impact of emerging technologies on work: a review of the evidence and implications for the human resource function. Emerald Open Res 1:5. https://doi.org/10.12688/emeraldopenres.12907.1
32. de Menezes LM, Kelliher C (2011) Flexible working and performance: a systematic review of the evidence for a business case. Int J Manag Rev 13:452–474. https://doi.org/10.1111/j.1468-2370.2011.00301.x
33. Fleming P (2017) The human capital Hoax: work, debt and insecurity in the era of uberization. Organ Stud 38:691–709. https://doi.org/10.1177/0170840616686129
34. World Bank (2019) Rankings & ease of doing business score: economy rankings. In: World Bank. https://www.doingbusiness.org/en/rankings. Accessed 30 May 2019
35. Boldrin M, Canova F, Pischke J-S, Puga D (2001) Inequality and convergence in Europe's regions: reconsidering European regional policies. Econ Policy 16:205–253. https://doi.org/10.2307/3601038
36. Sapir A, Aghion P, Bertola G et al (2004) An agenda for a growing Europe: making the EU economic system deliver. Oxford University Press, Oxford
37. Olczyk M (2014) Structural heterogeneity between EU 15 and 12 New EU Members—the Obstacle to Lisbon strategy implementation? Equilibrium 9:21–43. https://doi.org/10.12775/EQUIL.2014.023
38. Ederveen S, Groot HLF, Nahuis R (2006) Fertile soil for structural funds? A panel data analysis of the conditional effectiveness of European cohesion policy. Kyklos 59:17–42. https://doi.org/10.1111/j.1467-6435.2006.00318.x
39. Mendez C, Bachtler J (2011) Administrative reform and unintended consequences: an assessment of the EU Cohesion policy "audit explosion". J Eur Public Policy 18:746–765. https://doi.org/10.1080/13501763.2011.586802
40. Barbier J-C (2012) Tracing the fate of EU "social policy": changes in political discourse from the "Lisbon Strategy" to "Europe 2020". Int Labour Rev 151:377–399. https://doi.org/10.1111/j.1564-913X.2012.00154.x
41. Barca F (2008) An agenda for a reformed cohesion policy a place-based approach to meeting European Union challenges and expectations. Economics and Econometrics Research Institute, Brussels
42. Renda A (2014) The review of the Europe 2020 strategy: from austerity to prosperity?. Centre for European Policy Studies, Brussels
43. Bertolini P, Pagliacci F (2010) Lisbon strategy and EU countries' performance: social inclusion and sustainability. University of Modena and Reggio E., Faculty of Economics "Marco Biagi," Modena
44. Commission European (2015) Horizon 2020: first results. European Commission, Brussels
45. Button K (1998) Infrastructure investment, endogenous growth and economic convergence. Ann Reg Sci 32:145–162. https://doi.org/10.1007/s001680050067

46. Fernald JG (1999) Roads to prosperity? Assessing the link between public capital and productivity. Am Econ Rev 89:619–638. https://doi.org/10.1257/aer.89.3.619
47. Crescenzi R, Rodríguez-Pose A (2012) Infrastructure and regional growth in the European Union*. Pap Reg Sci 91:487–513. https://doi.org/10.1111/j.1435-5957.2012.00439.x
48. Vlahinić Lenz N, Pavlić Skender H, Mirković PA (2018) The macroeconomic effects of transport infrastructure on economic growth: The case of Central and Eastern E.U. member states. Econ Res Istraživanja 31:1953–1964. https://doi.org/10.1080/1331677X.2018.1523740
49. Baños-Pino J, Coto-Millán P, Rodríguez-Álvarez A (1999) Allocative efficiency and over-capitalization: an application. Int J Transp Econ 26:181–199. https://doi.org/10.2307/42747743
50. Schaap R, Richter A (2019) Overcapitalization and social norms of cooperation in a small-scale fishery. Ecol Econ 166:106438. https://doi.org/10.1016/j.ecolecon.2019.106438
51. Peng T-C (2011) Overcapitalization and cost escalation in housing renovation. New Zeal Econ Pap 45:119–138. https://doi.org/10.1080/00779954.2011.556074
52. Anand A, Felman J, Sharma N, Subramanian A (2015) Paranoia or Prudence? Econ Polit Wkly 53:7–8
53. Gorzelak G (2014) Wykorzystanie środków Unii Europejskiej dla rozwoju kraju – wstępne analizy. Stud Reg i Lokal 3:5–25. https://doi.org/10.7366/1509499535701
54. Komornicki T (2013) Work package 6: territorial dimension of EU integration as challenges for cohesion policy, Task 4. Assessment of infrastructure construction, its role in regional development. GRINCOH Project, Warsaw
55. Ford R, Goodwin M (2017) Britain after Brexit: a nation divided. J Democr 28:17–30
56. Koca BT (2019) Bordering practices across Europe: The rise of "Walls" and "Fences." Migr Lett 16:183–194. https://doi.org/10.33182//ml.v16i2.634
57. Baldwin-Edwards M, Blitz BK, Crawley H (2019) The politics of evidence-based policy in Europe's 'migration crisis'. J Ethn Migr Stud 45:2139–2155. https://doi.org/10.1080/1369183X.2018.1468307
58. Ivanova D (2016) Migrant crisis and the Visegrád group's policy. Int Conf Knowl Organ 22:35–39. https://doi.org/10.1515/kbo-2016-0007
59. Estevens J (2018) Migration crisis in the EU: developing a framework for analysis of national security and defence strategies. Comp Migr Stud 6:28. https://doi.org/10.1186/s40878-018-0093-3

Evaluating Green Infrastructure via Unmanned Aerial Systems and Optical Imagery Indices

Matjaž N. Perc and Giuseppe T. Cirella

Abstract Within small-sized areas, one of the most time and cost-effective remote sensing techniques for evaluating green infrastructure is unmanned aerial systems (UAS) in conjunction with optical imagery indices. In terms of urban sustainability, this approach can be applicable when, otherwise, expensive or up-to-date imagery is unavailable. This chapter illustrates the use of UAS as an adjunct to best urban planning practice and landscape infrastructural design and pieces together an environmental application by using flood mitigation as an example. The study area, located in Slovenia, assembles bottom-up imagery to produce a UAS-based flood map from predicted and potential rainfall. The basis of this research is to utilize simulation data from climate scenarios to show the intensification of precipitation change with pronounced change in seasonal variability. As such, green urban infrastructure mitigation policy can utilize UAS as a complimentary environmental monitoring tool for relating natural and urban systems.

Keywords Urban green areas · Green infrastructure · Remote sensing · Unmanned aerial vehicles · Drones · Slovenia

1 Introduction

Predicting climatic variability undoubtedly is a challenging factor for the current meteorological and weather prediction systems [1–3]. One important aspect in examining this variability includes water level via seasonal change and precipitation. Levels of water scarcity that result in pronounced changes with decreased level of precipitation in the summer and increased in winter are especially alarming [4]. As

M. N. Perc (✉)
Faculty of Civil Engineering, Transportation Engineering and Architecture,
University of Maribor, Maribor, Slovenia
e-mail: matjaz.nekrep@um.si

G. T. Cirella
Faculty of Economics, University of Gdansk, Sopot, Poland
e-mail: gt.cirella@ug.edu.pl

© Springer Nature Singapore Pte Ltd. 2020
G. T. Cirella (ed.), *Sustainable Human–Nature Relations*,
Advances in 21st Century Human Settlements,
https://doi.org/10.1007/978-981-15-3049-4_9

such, urban green areas and green infrastructure are becoming of greater importance to this phenomenon [5]. As an adjunct to best urban planning practice and landscape infrastructural design, the use of unmanned aerial systems (UAS) and red–green–blue (RGB) optical imagery utilized in combination with geographic information systems (GIS) can be used to detect change in urban green areas [6] as well as calculate and compare different RGB-based indices for that purpose [7–11].

Applications in sharing information that make urban forests, green infrastructure, and woody plants thrive are no longer restricted to the academic sphere of study. With the price of UAS (i.e., drone) technology becoming much more affordable, applied research into urban green infrastructure has started moving into other remote sensing-related systems for detecting change in urban forests, ameliorating the urban heat island, early detection of pests, and disease in woody plant crops [9, 11]. Also, a number of mitigation-related concerns including trees and solar collectors, improving green infrastructure design, planning and making urban policy (e.g., for sidewalks and roadside design for trees), integrating green infrastructure, and designing compact urban settings (e.g., improving parking lots and vertical green building) play an important, contributive role for best urban landscape practice.

This chapter utilizes UAS technology in conjunction with optical imagery indices to best show how predicted changes in climate extremes, among others, point to substantial temperature extremes in an urban setting. The chapter focalizes on the application-end of the research and explicitly uses imagery enhancement to examine the likelihood and frequency of heavy precipitation or the proportion of total rainfall from heavy falls that can be correlative with twenty-first-century estimates [12, 13]. As a working example, we utilize simulation data from climate scenarios to show a significant increase in the average annual temperature—seasonally—and the intensification of precipitation changes with pronounced changes in seasonal variability [14]. The challenges associated with extreme meteorological events are numerous and complex, as such, the example of aging urban infrastructure commonly unprepared for extreme storm events is considered as residual to approach [15]. Green urban infrastructure based on existing green areas is one answer to those challenges [16]. A breakdown of the chapter utilizes a part of Slovenia and illustrates stage-developed imagery of spatial results using UAS to show conditions in an urban catchment and potential for flood.

2 Case Study

The study area is in the southwestern part of Maribor, Slovenia, specifically within the Radvanje District which lies under the Pohorje ski slopes. This area is important due to an increased level of urbanization that is mostly associated with the district's prestigious environmental backdrop and economic development. We utilize UAS orthophotography and geometrically corrected scaling (i.e., ortho-rectified the imagery) to create a uniform, true representation of the study site (Fig. 1).

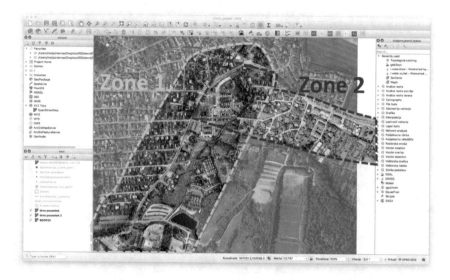

Fig. 1 Radvanje District, Maribor, Slovenia; created using GIS in combination with UAS orthophotography

To help piece together the district's likelihood and frequency of heavy precipitation or the proportion of total rainfall from heavy falls as well as probability of runoff damage (i.e., from flooding), we examined the drainage system and utilized the historical alluvial record for Slovenia as a whole [13, 17]. The urban area of Radvanje District is covered by an existing mixed system drainage network which originally was intended for a much smaller population with much less paved and cementification. Recently, extreme weather events have had an alarming impact on the area. Stormwater that cannot permeate the urbanized setting (e.g., roads and parking lots) have recorded an augmentation in flooding and infrastructural damage. The geographic landscape of the district includes a small river reach in the middle of the site which is highly and inappropriately regulated as well as improperly used to reduce runoff flow volume from the drainage system; conversely, its capacity is limited and causes additional flooding downstream when extremely heavy rainfall occur. A mapped layout of the district's existing mixed system drainage network was created using research from the Faculty of Civil Engineering, Traffic Engineering, and Architecture in Maribor and adapted from Jecl et al.'s [18] research (Fig. 2).

Fig. 2 Scheme of mixed system drainage network for the study area [18]

3   Unmanned Aerial Systems as a Source of Instant Environmental Data

3.1   Why Use Unmanned Aerial Systems?

The utility of UAS (i.e., drone) technology as an optimal source of spatial data, especially in the case of semi-size areas (e.g., the study area), have proven to be a viable method of attaining environmental information. On the one hand, UAS can help fill in the gap between digital remote sensing and aerial data while on the other provide common terrestrial geodetic survey information. As noted, the role of drone producers and consumers has begun to blur and merge. Today, the prosumer drone which falls between consumer and commercial-grade standards in quality, complexity, and functionality offers excellent results under certain restrictions (e.g., ground control points (GCPs), permissions, and weather conditions). They are efficient in terms of the amount of time it takes to produce imagery as well as highly affordable. A price breakdown of the different types of UAS shows the versatility of the technology and its market capacity (Fig. 3).

Options for using UAS for environmental phenomena are continually improving, especially in recent years. Air and water sensors that can measure pollutants and subsets of the electromagnetic spectrum have become smaller, cheaper, and more bundled into comprehensive units [20]. According to Gallacher [21], "aerial sensor

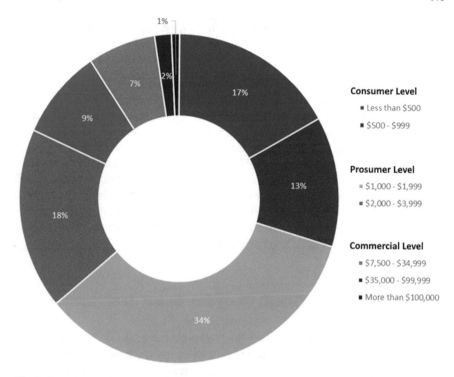

Fig. 3 Purchase price of UAS (USD). Adopted from Skylogic Research, LLC [19]

platforms have also expanded in the form of low-altitude unmanned aerial vehicles (i.e., micro-drones), but their use in populated spaces is increasingly restricted for safety and privacy reasons." Potential applications of UAS for environmental monitoring and management of urban spaces include platform-based aerial sensing (i.e., electromagnetic spectrum, atmospheric composition, data collection from detached sensors, and aerial sensing of sound) and aerial transport. The use of RGB-based indices simulates the electromagnetic spectrum (i.e., visible light) which can offer much greater detail with less atmospheric interference. Gallacher [21] states UAS "will normally require [...] development to automate the analysis of the enormous amount of data produced." The use of UAS-based services must outweigh the risks and also be competitive with ground-based methods before the benefits and risk can be widely adopted and subsequently utilized [21].

3.2   Green Area Detection Using Unmanned Aerial Systems Visible Imagery

By surveying the study area using unmanned aerial vehicles (UAVs) that capture photographs via a preplanned autonomous flight and processing visible spectrum orthophotographs, we can obtain a number of vegetation indices (i.e., Visible Atmospheric Resistant Index, Triangular Greenness Index, and Green Leaf Index) from the visual RGB data [22, 23]. Using the collected data, we combine color bands from the RGB images and create graphical indicators that separate green areas (i.e., woods, meadows, and gardens) from the built environment (i.e., buildings, roads, and pavement areas). As such, green area detention is vital in demonstrating observed vegetation mapping. The present study compares the utility of UAV images specifically in terms of image quality, efficiency, and classification accuracy. Both object- and pixel-based classification approaches are integrated.

3.3   Types of Spatial Results Using Unmanned Aerial Systems

Structure from motion (SfM) is an excellent and rapid approach to surveying existing site conditions [24]. It allows the user to safely and quickly survey existing terrain, infrastructure, buildings, water bodies, and green area and accurately model a site. To date, it is also one of the most cost-effective solutions. This new technology allowed us to survey the study site, using UAVs to capture photographs, by eliminating the need for on-site surveys and creating an accurate model (i.e., with texturized mesh) at an engineered standard. The preplanned and autonomous flight was inputted to cover the whole study area in a systemic back-and-forth manner. The study area was subdivided into two zones. An example of the spatial results from Zone 2 can be seen in Fig. 4. Using SfM, photogrammetric range imaging technique was used to piece together three-dimensional structures from two-dimensional image sequences. Computer vision and visual perception are used to tie the imagery together and overlay it in a GIS. SfM phenomenon is projected via the retinal motion field of the moving UAV.

From the UAV imagery, a digital surface model (DSM) in conjunction with a digital elevation model (DEM) was created. Piecing together the DSM and DEM, we excluded vegetation and buildings to create a digital terrain model (DTM) (Fig. 5). Moreover, with the DTM, a georeferenced, detailed orthomosaic (i.e., using the GeoTif data) inputted into a three-dimensional sequence was performed to create a three-dimensional model. The process involved an "on the fly" fly-process-analysis as seen in Fig. 6.

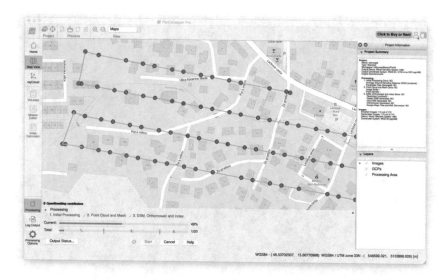

Fig. 4 UAV flight plan with the location points of taken photos from Zone 2

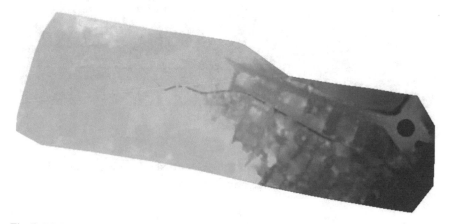

Fig. 5 Digital terrain model of Zone 2

4 Results: Case Study of the Unmanned Aerial Systems Survey

Using a UAV, we captured 446 aerial photographs in a mere 25 min and incorporated base imagery from four perimeter GCPs. The use of GCPs is commonly used for georeferencing in remote sensing and UAS. They help with careful surveying what would otherwise be a time-consuming and labor-intensive and excessively costly task [25, 26]. Using CGPs, we detailed the surveyed area of 43 ha from the height

Fig. 6 Three-dimensional model of the study site

of 90 m with an estimated accuracy of 2.7 cm per pixel. In doing so we calculated the Green Leaf Index by using an algorithmic formula (Eq. 1).

$$GLI = (2 \times G - R - B)/(2 \times G + R + B) \tag{1}$$

where GLI = Green Leaf Index, G = green, R = red, and B = blue.

The Green Leaf Index for the two zones is illustrated in Fig. 7 (i.e., Zone 1) and Fig. 8 (i.e., Zone 2). To further distinguish the developed areas, living plants (e.g., vegetation and wheat) are differentiated from soil and non-living matter [27]. The resulting imagery used semi-natural (e.g., urban forest) and agricultural areas (e.g., grassland) to estimate the Green Leaf Index.

Utilizing the Green Leaf Index data, secondary imagery manipulation—via segmentation—was performed [28]. Sample images at quadrat 10 (Q10) of the study site illustrate urban green areas (Fig. 9) contrasted with pavement areas (Fig. 10). Segmented mapping can be utilized as an urban green indicator and show urbanization trends as well as factors relating to green urban sustainability.

Fig. 7 Map of Green Leaf Index of Zone 1

Fig. 8 Map of Green Leaf Index of Zone 2

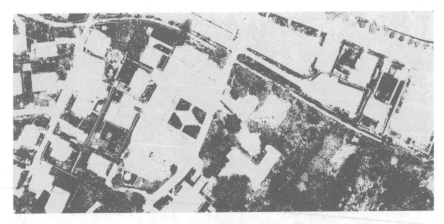

Fig. 9 Urban green areas (i.e., colored green) at Q10, Leaf Area Index = 41.34%

Fig. 10 Pavement areas (i.e., colored black) a Q10, visible ground area = 58.66%

5 Flood Map

The use of UAS to collect data from the whole of the Radvanje District was relatively quickly undertaken. Based on the data acquired, green and pavement areas
can be used as a base for future planning for green infrastructure area wide. Sustainable urban green design and planning can utilize this information for a number of
environmental management tasks. In respect to flood mitigation, we calculated the
amount of stormwater, adapted from the Slovenian Environment Agency, and used
municipal GIS data overlaid with the existing drainage network to create a broad-
ened representation of the area's flood potential. With extensive UAS data gathering,
precise spatial data was acquired and analyzed within the context of RGB imagery
results.

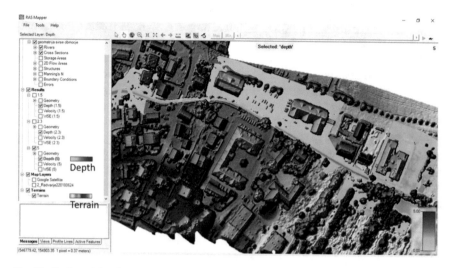

Fig. 11 Projected flood map at Q10

This approach has been proven useful by incorporating green and pavement areas to calculate runoff coefficient [29]. As such, a flood area map at quadrat 10 (Q10) of the study site illustrates the failure of the current urban catchment [5] and models stormwater runoff (Fig. 11). The results show a realistic and worrisome future for the Radvanje District if proper mitigation is not implemented. Results indicate severe flooding along the main river with northerly lower land confronting flood depth of up to 4 m.

6 Conclusion

Evaluating green infrastructure via UAS and optical imagery indices can be a very time and cost-effective manner of assessing a small-sized area. In terms of urban sustainability, this approach can be an effective and important way of acquiring otherwise expensive or up-to-date imagery from other more traditional aerial platforms. One of the most striking features of UAS data is its high spatial resolution which can be scaled down to as small as a centimeter, or potentially even finer, as is demonstrated by much of the current research [30, 31]. Scientists, and the greater community, are only just beginning to explore remote sensing at this scale [32].

In terms of facilitating better mitigation strategies for environmental applications, our example of constructing a bottom-up design for Radvanje District's flood zone illustrates the potential for UAS technology and its benefit for sustainable planning. Specific to the study site, future research could incorporate impact from distributed retention and infiltration measures on urban runoff [33] which would add an additional layer of information and assist with best infrastructure design. Measures such as sustainable urban drainage systems should be best designed to reduce the quantity of water that would otherwise overflow from the combined sewer system (i.e., into the recipient). UAS remote sensing data can play an important part in contributing to quality flood control and prevention [34]. This approach can contribute to a best practices policy as well as community awareness and need to mitigate infrastructural damage, save lives, and build preventative, strong community awareness. Enhanced decision making and accelerated stakeholder buy-in for explanations and solutions are other potential uses [35]. At length, environmental monitoring via UAS can play a significant role in diagnosing climate and management-level impact via natural and urban systems. Enhancing the understanding of all environmental phenomena by way of UAS-based tools can optimize allocation and distribution of resources, assessment and forecasting of meteorological patterns, and mitigate ground-based measurements that radically improves environmental monitoring.

References

1. Mahmood R, Jia S, Zhu W (2019) Analysis of climate variability, trends, and prediction in the most active parts of the Lake Chad basin, Africa. Sci Rep 9:6317. https://doi.org/10.1038/s41598-019-42811-9
2. Butler CD (2018) Climate change, health and existential risks to civilization: a comprehensive review (1989–2013). Int J Environ Res Public Health 15. https://doi.org/10.3390/ijerph15102266
3. Stockdale TN, Alves O, Boer G et al (2010) Understanding and predicting seasonal-to-interannual climate variability—the producer perspective. Procedia Environ Sci 1:55–80. https://doi.org/10.1016/J.PROENV.2010.09.006
4. Hanel M, Rakovec O, Markonis Y et al (2018) Revisiting the recent European droughts from a long-term perspective. Sci Rep 8:9499. https://doi.org/10.1038/s41598-018-27464-4
5. Tao J, Li Z, Peng X, Ying G (2017) Quantitative analysis of impact of green stormwater infrastructures on combined sewer overflow control and urban flooding control. Front Environ Sci Eng 11:1–12. https://doi.org/10.1007/s11783-017-0952-4
6. Mckinnon T, Hoff P (2017) Comparing RGB-based vegetation indices with NDVI for drone based agricultural sensing
7. Lisein J, Michez A, Claessens H, Lejeune P (2015) Discrimination of deciduous tree species from time series of unmanned aerial system imagery. PLoS ONE 10:e0141006. https://doi.org/10.1371/journal.pone.0141006
8. Ballesteros R, Ortega JF, Hernandez D, Moreno MA (2018) Onion biomass monitoring using UAV-based RGB imaging. Precis Agric 19:840–857. https://doi.org/10.1007/s11119-018-9560-y
9. Themistocleous K (2019) DEM modeling using RGB-based vegetation indices from UAV images. In: Papadavid G, Themistocleous K, Michaelides S et al (eds) Seventh international conference on remote sensing and geoinformation of the environment (RSCy2019). SPIE, p 21
10. Goodbody TR, Coops NC, Hermosilla T et al (2018) Assessing the status of forest regeneration using digital aerial photogrammetry and unmanned aerial systems. Int J Remote Sens 39:5246–5264. https://doi.org/10.1080/01431161.2017.1402387
11. Zheng H, Cheng T, Li D et al (2018) Evaluation of RGB, color-infrared and multispectral images acquired from unmanned aerial systems for the estimation of nitrogen accumulation in rice. Remote Sens 10:824. https://doi.org/10.3390/rs10060824

12. Field CB, Barros V, Stocker TF et al (2012) Managing the risks of extreme events and disasters to advance climate change adaptation: special report of the intergovernmental panel on climate change
13. Trobec T (2017) Frequency and seasonality of flash floods in Slovenia. Geogr Pannonica 21:198–211. https://doi.org/10.5937/gp21-16074
14. Bertalanič R, Dolinar M, Ključevšek N et al (2017) Ocena podnebnih sprememb v Sloveniji do konca 21: Stoletja Povzetek temperaturnih in padavinskih povprečij. Slovenvian Environment Agency, Ljubljana
15. Skougaard Kaspersen P, Høegh Ravn N, Arnbjerg-Nielsen K et al (2017) Comparison of the impacts of urban development and climate change on exposing European cities to pluvial flooding. Hydrol Earth Syst Sci. https://doi.org/10.5194/hess-21-4131-2017
16. Ramos H, Pérez-Sánchez M, Franco A, López-Jiménez P (2017) Urban floods adaptation and sustainable drainage measures. Fluids. https://doi.org/10.3390/fluids2040061
17. Bezak N, Horvat A, Šraj M (2015) Analysis of flood events in Slovenian streams. J Hydrol Hydromech 63:134–144. https://doi.org/10.1515/johh-2015-0014
18. Jecl R, Kramer Stajnko J, Nekrep-Perc M, Grajfoner B (2018) Študija prevodnosti kanalizaci-jskega sistema zgornje radvanje – kanal radvanjski trg. Fakulteta za gradbeništvo, prometno inženirstvo in arhitekturo, Maribor
19. Skylogic Research (2018) Skylogic Research, LLC. In: San Fr. Bay Area, Silicon Val. http://droneanalyst.com/. Accessed 22 Jul 2019
20. Ruwaimana M, Satyanarayana B, Otero V et al (2018) The advantages of using drones over space-borne imagery in the mapping of mangrove forests. PLoS ONE 13:e0200288. https://doi.org/10.1371/journal.pone.0200288
21. Gallacher D (2017) Drone applications for environmental management in urban spaces: a review. Int J Sustain L Use Urban Plan 3. https://doi.org/10.24102/ijslup.v3i4.738
22. Kang Y, Özdoğan M, Zipper SC et al (2016) How universal is the relationship between remotely sensed vegetation indices and crop leaf area index? A global assessment. Remote Sens 8. https://doi.org/10.3390/rs8070597
23. Mckinnon T (2017) Comparing RGB-based vegetation indices with NDVI For drone based agricultural sensing. In: Agribotix LLC, Boulder, CO, USA. https://agribotix.com/wp-content/uploads/2017/05/Agribotix-VARI-TGI-Study.pdf. Accessed 28 Jul 2019
24. Nesbit P, Hugenholtz C (2019) Enhancing UAV–SfM 3D model accuracy in high-relief landscapes by incorporating oblique images. Remote Sens 11:239. https://doi.org/10.3390/rs11030239
25. Han X, Thomasson J, Xiang Y et al (2019) Multifunctional ground control points with a wireless network for communication with a UAV. Sensors 19:2852. https://doi.org/10.3390/s19132852
26. Nguyen T (2015) Optimal ground control points for geometric correction using genetic algorithm with global accuracy. Eur J Remote Sens 48:101–120. https://doi.org/10.5721/EuJRS20154807
27. Liu X, Wang L (2018) Feasibility of using consumer-grade unmanned aerial vehicles to estimate leaf area index in Mangrove forest. Remote Sens Lett 9:1040–1049. https://doi.org/10.1080/2150704X.2018.1504339
28. Wardley NW, Curran PJ (2007) The estimation of green-leaf-area index from remotely sensed airborne multispectral scanner data. Int J Remote Sens 5:671–679. https://doi.org/10.1080/01431168408948850
29. Lee JG, Nietch CT, Panguluri S (2018) Drainage area characterization for evaluating green infrastructure using the storm water management model. Hydrol Earth Syst Sci 22:2615–2635. https://doi.org/10.5194/hess-22-2615-2018
30. Manfreda S, McCabe M, Miller P et al (2018) On the use of unmanned aerial systems for environmental monitoring. Remote Sens 10:641. https://doi.org/10.3390/rs10040641
31. Yang G (2018) High resolution satellite imaging sensors for precision agriculture. Front Agric Sci Eng 0:0. https://doi.org/10.15302/J-FASE-2018226
32. Simic Milas A, Sousa JJ, Warner TA et al (2018) Unmanned aerial systems (UAS) for environmental applications special issue preface. Int J Remote Sens 39:4845–4851. https://doi.org/10.1080/01431161.2018.1491518

33. Jones S, Somper C (2014) The role of green infrastructure in climate change adaptation in London. Geogr J 180:191–196. https://doi.org/10.1111/geoj.12059
34. Radinja M, Banovec P, Matas JC, Atanasova N (2017) Modelling and evaluating impacts of distributed retention and infiltration measures on urban runoff. Acta Hydrotech 30:51–64
35. Colomina I, Molina P (2014) Unmanned aerial systems for photogrammetry and remote sensing: a review. ISPRS J. Photogramm, Remote Sens, p 92

Urbanization and Sustainability Strategies

Urban Sustainability: Integrating Ecology in City Design and Planning

Alessio Russo and Giuseppe T. Cirella

Abstract Urban sustainability depends on ecosystem services and biodiversity which directly affects quality of urban life. At present, urbanization is having a drastic effect on the way human beings interact with the world around us. Urbanized environments tend to lessen the amount of habitat and increase habitat fragmentation. This important factor stresses the need for sound urban sustainability thinking as well as related urban planning and urban design processes. Adaptive urban knowhow is as the root of this chapter in which a number of exploratory concepts and notions are put forth with the intention of creating dialogue between ecosystem services and human well-being (i.e., through concerted ecological, economic, and social action). The chapter begins with a look at urban sustainability, explores sustainable urban strategies, considers a number of ideas under the umbrella of urban green infrastructure—reviewing a number of case examples—and concludes with background research in properly developing sustainable models and tools. Integrating ecology in city design and planning should support resilience-orient development and highlight a synergetic, evolutionary form of multi-disciplinary sustainability.

Keywords Nature-based solutions · Ecological corridors · Green infrastructure · Urban ecology

1 Introduction

Humanity has become a major force of nature in this new Anthropocene epoch, and we continue to deploy the biosphere with increasing intensity across all ecosystems to withstand exponential population growth and meet the demand for natural resources [1]. To date, urbanization, an important factor in the worsening of air quality [2],

A. Russo (✉)
School of Arts, University of Gloucestershire, Cheltenham, UK
e-mail: arusso@glos.ac.uk

G. T. Cirella
Faculty of Economics, University of Gdansk, Sopot, Poland
e-mail: gt.cirella@ug.edu.pl

© Springer Nature Singapore Pte Ltd. 2020
G. T. Cirella (ed.), *Sustainable Human–Nature Relations*,
Advances in 21st Century Human Settlements,
https://doi.org/10.1007/978-981-15-3049-4_10

has a weighted impact on the loss of the world's biodiversity and homogenization of its biota through the replacement of non-urban specialist species [3, 4]. It tends to decrease the amount of habitat and increase habitat fragmentation at the same time, while the relationship between habitat loss and habitat fragmentation, due to the urbanization phenomena, is generally "monotonic-linear, exponential, or logarithmic—[indicating that the degree of] habitat fragmentation per se increases with habitat loss" [5]. Increased impervious surfaces resulting from urbanization modifies the water cycle by reducing the infiltration of stormwater and increasing surface runoff. Even more dramatically, these impermeable surfaces contribute to increased urban flood events [6].

Achieving urban sustainability via urban planning and urban design requires complex goal-oriented decision-making processes, which consider interactions among many factors, such as economics, social and cultural context, physical environment, and ecology [7]. Urban sustainability is an "adaptive process of facilitating and maintaining a virtuous cycle between ecosystem services and human well-being through concerted ecological, economic, and social actions in response to changes within and beyond the urban landscape" [8]. In the last quarter century, the discipline of urban ecology that contributes to urban sustainability evolved from an approach that focuses on ecological structures, functioning within a more holistic manner, where the city itself is the ecosystem and human beings are the system's dominant species [9]. Within this integrated social-ecological system [10], the city is dependent on ecosystem goods and services from natural ecosystems to maintain sustainability and depend on human management and intervention to maintain viable levels of sustainable functionality [11, 12]. For the renewal of urban areas, landscape urbanists suggest that landscapes should replace buildings and transportation systems as the principal organizing structure in urban design [13]. On the other hand, traditional urbanism and new urbanism relegate green areas to places that are too expensive to develop, while landscape urbanism theorists advocate the integration of ecology in city design and planning [13]. This chapter divulges into the development processes of urban sustainability and examines the integration of ecology and green spaces in city design and planning. In doing so, there is a perception that increased tree cover and density in urban contexts are associated with increased rates of criminality even though several studies concluded the opposite [14]. For instance, in Baltimore, Troy et al. [15] found that a 10% increase in tree canopy was associated with roughly a 12% decrease in crime, while Escobedo et al. [16] discovered the amount of public green areas was not significantly related to homicide occurrence in Bogota, Colombia. The scope of this chapter expands discussional notions of ecological concepts and approaches for the design and planning of sustainable cities by furthering the knowledge-base of existing tools and models for measuring, modeling, and valuing city-ecosystem services. It explores urban green infrastructure, urban ecological corridors, urban biosphere reserves and reviews case study examples.

2   Urban Sustainability

2.1   *Ecological Planning, Ecosystem Services, and Eco-cities*

The evolution of ecological planning can be traced back to the early works of Frederick Law Olmsted Sr. and Jr., Sir Ebenezer Howard, Frank Lloyd Wright, Patrick Geddes, Lewis Mumford, and Ian McHarg [6, 17]. Over the last decade, green infrastructure—in connotation with sustainability—has risen as a disciplinary field in which significant interest has been focused on the city and regional planning; however, this discussion is not entirely valid. Significant elements can be dated back to the work of Frederick Law Olmsted Sr., in the nineteenth century, in which rooted work in urban planning and landscape architecture professions first became evident [17]. Olmsted's work within contemporary green infrastructure theory and practice included ecosystem services and human well-being [17]. In the late nineteenth century, the first inspirations of the garden city movement were put forth by Sir Ebenezer Howard [18] when he wrote "To-morrow: A peaceful path to real reform." Howard notes Henry George's [19], "Progress and poverty: An inquiry into the cause of industrial depressions and of increase of want with increase of wealth: The remedy," in which initial attempts are put together to recreate English village life by bringing "green" back into towns and by controlling urban size and growth. The garden city movement was later replaced by the concept of the "Abercrombie Plan" [20] which was followed by the idea of Jacobs' [21] "Living City," the idea that helped formulate the concept of Disneyland. In France, ideas from the Swiss–French architect Charles-Édouard Jeanneret, known as Le Corbusier, established the "modernist" city during the 1920s and 1930s, which came to be internationally renowned and, to date, still shapes urban planning in many parts of the world. Le Corbusier held that the ideal city was tidy, well-ordered and highly controlled; cities should replace wasted space with efficient transportation corridors and residential tower blocks that separated "flowing" mono-functional zones of open space [22].

In the early twentieth century, American architect Frank Lloyd Wright promoted the ideal city in terms of low density, dispersed urban forms (i.e., each family having their own plot) [23]. Since the 1990s, a number of cities have created new neighborhoods taking environmental factors into consideration. Shanghai announced plans to build the city of the future on an island at the mouth of China's Yangtze River, and in the same way, Singapore has planned new ecological neighborhoods [24]. There is a growing interest in sustainable concepts such as the eco-city (i.e., including the sustainable city, smart city, low-carbon city, and resilient city) especially since the passing of the new millennium in which significant global proliferation of diverse policies and practical initiatives have come to the forefront [25]. One example is the European Eco-Viikki project established in Finland as an ecological residential area built between 1999 and 2004. As a planned competition, organized by the City of Helsinki and the Eco-Community Project, challenging planners to envision ecological views and integrate model solutions at the neighborhood level [26] (Fig. 10.1).

Fig. 10.1 Eco-Viikki, Finland—(left) ecological building and "green fingers" with gardening plots, (right) green roof and greenhouse. Photographs taken by A. Russo on August 15, 2012

The planning stages of urbanization lead us to explore urban strategies and scholarship and the how the rethinking of urban design can mitigate unsustainable action and promote sound development.

2.2 Sustainable Urban Strategies and Scholarship

As the world experiences unprecedented urbanization—presenting extreme challenges for contemporary urban planners, designers, and policy-makers—sustainable urban strategies and an expert level of ecological scholarship are needed. It will require boosting urban models and strategies capable of generating new knowledge and improving practical skills [27]. It is fair-minded to consider we are already living in the Age of Cities, characterized by increasing population and urban centrality in the achievement or declination of itself—reflective of region, country, or the global economy as a whole [28]. Sustainable and resilient cities have become the world's leading urban development paradigm [29]. Recently, Zhang and Li [30] highlighted the difference between urban sustainability and urban resilience, pointing out that rational urban development can only be reached when it is both sustainable and resilient. Moreover, they define urban sustainability as "the active process of synergetic integration and co-evolution between the subsystems making up a city without compromising the possibilities for the development of surrounding areas and contributing by this means toward reducing the harmful effects of development on the biosphere" [30]. Urban resilience, on the other hand, as "the passive process of monitoring, facilitating, maintaining, and recovering a virtual cycle between ecosystem services and human well-being through concerted effort under external influencing factors" [30].

The concept of urban sustainability is an inclusive characteristic of urban resilience and sustainable development needs as well as the inclusivity of how to develop resilience-oriented strategies [31]. Several critical scholars, however, have warned that resilience theories fail to tackle equity, justice, and power [32]. The official political implementation of the sustainable city started in 1972 when the significance of creating sustainable urbanization models first came to light, internationally, at the 2003 United Nations Conference on Human Environment in Stockholm [29]. According to Principle 16 of the Declaration from that conference "planning must be applied to human settlements and urbanization with a view to avoiding adverse effects on the environment and obtaining maximum social, economic, and environmental benefits for all" [33]. This principle has been strengthened by the Sustainable Development Goal 11, which aims to "make cities inclusive, safe, resilient, and sustainable" by 2030 [34].

Lately, cities and towns from around the world have been consciously encouraged to be smarter (and even smarter) and therefore more sustainable by creating and applying big data systems and applications across different metropolitan areas in the hope of achieving the necessary level of sustainability and enhancement in living standards [35]. According to the World Bank [36], "smart cities make urbanization more inclusive, bringing together formal and informal sectors, connecting urban cores with peripheries, delivering services for the rich and the poor alike, and integrating the migrants and the poor into the city. Promoting smart cities is about rethinking cities as inclusive, integrated, and livable." The application of information and communications technology (ICT) in cities can produce various benefits (e.g., reducing resource consumption, improving the utilization of existing infrastructure capacity, making new services available to citizens and commuters, and improving commercial enterprises) [37]. Specifically, the ICT smart city agenda is designed to address a wide range of urban problems (i.e., cities can increase rates of economic growth, competitiveness, and innovation while reaching sustainable goals such as emissions reduction, enhancing energy effectiveness, and enhancing quality of life) [38, 39]. Zhu et al. [40] recently unveiled potential connections between urban smartness and resilience in China, showing important findings that correlate to two. Cities can be both smart and sustainable, which require the preservation of the ecological equilibrium and social identity of urban communities embedded in tangible and intangible heritage while promoting creativity and technology in order to increase their productivity and resilience and thus improve the welfare and quality of its citizenry [41]. The United Nations Educational, Scientific and Cultural Organization (UNESCO) [41] provides several recommendations supporting sustainable urbanization by encompassing environmental, social, and economic standpoints. Its recommendations can be used by policy-makers and urban planners alike; however, there is a lack of sustainable urban development indicators which makes it difficult for urban sustainability to be properly evaluated in a number of situations. In particular, those with high sensitive, fragile ecology, and lower threshold limits [42] require a more comprehensive list of indicators and range of weightages varying according to different city-contexts. By rethinking urban design, architecture, landscape design, urban transport, and planning, we can transform our cities and urban landscapes into "urban ecosystems"—a forefront notion for mitigating climate change (e.g., sustainable transportation and clean energy) as well as urban adaptation solutions (e.g., floating houses and nature-based solutions) [43].

2.3 Urban Green Infrastructure

Urban green infrastructure is a "hybrid infrastructure of green spaces and built systems (e.g., urban forests, wetlands, parks, green roofs, and walls) that together can contribute to ecosystem resilience and human benefits through ecosystem services" [44]. Urban green infrastructure has been documented as providing several economic, aesthetic, cultural, and architectural benefits [44]. In particular, it provides multiple ecosystem services and goods such as carbon storage and sequestration, water and air purification as well as offset air pollution and emissions from cities. Urban green infrastructure can reduce mortality from heatwaves and improve human thermal comfort through the altering of albedo of surfaces as well as cooling of atmospheric temperatures through shading and evapotranspiration [44–48]. There is increasing evidence that contacts with nature can make a significant contribution to human health and well-being [5, 49–51]. This is because natural ecosystems provide a variety of services some of which promote basic human needs (e.g., limiting the spread of disease or reducing air contaminants) [52]. Furthermore, several systematic reviews and meta-analyses pooling data have confirmed that exposure to urban green infrastructure results in the reduction of all-cause mortality, particularly cardiopulmonary and cancer mortality [53]. Edible green infrastructure and nature-based solutions are complementary notions within the urban green infrastructural discourse and merit further discussion.

2.3.1   Edible Green Infrastructure

Food insecurity is an important health problem and an underrecognized social determinant of health [54]. Edible green infrastructure is a novel approach that can improve resilience and quality of life in cities and boost food insurance [55]. It is a sustainable planned network of edible food components and structures, within the urban ecosystem that can manage and design the provisioning of ecosystem services. Illustrations of edible green infrastructure are shown in Fig. 10.2.

Edible green infrastructure typologies are provided and macrocategorized in conjunction with eight sub-classifications of urban agriculture: (1) edible urban forests and edible urban greening, (2) edible forest gardens, (3) historic gardens and parks and botanic gardens, (4) school gardens, (5) allotment gardens and community gardens, (6) domestic and home gardens, (7) edible green roofs and vegetable rain gardens, and (8) edible green walls and facades [55]. An edible green infrastructure approach can interplay environmentally, socially, and economically to urban sustainability and food security. As noted, the typologies of edible green infrastructure can

Fig. 10.2 Edible green infrastructure, (left) Berlin, Germany and (right) Vancouver, Canada; left photograph taken by A. Russo on 20 June 2018, right photography taken by G. T. Cirella on September 19, 2018

vary significantly; however, additional factors such as ecosystem disservices (e.g., potential health risks caused by heavy metals and organic chemical contaminants often found throughout urban settings) must be considered. There are a number of recommendations and guidelines for incorporating edible green infrastructure into urban planning and design [56] that should be considered for future research.

2.3.2 Nature-Based Solutions

Nature-based solutions is a new term in environmental research, planning, and management [57]. The nature-based solutions concept is closely related to other modern approaches including sustainability, urban resilience, ecosystem services, ecosystem-based adaptation, coupled human and environment, and green–blue infrastructure; however, "nature-based solutions represent a more efficient and cost-effective approach to development than traditional approaches" [58]. At the core, "they are solutions to societal challenges that are inspired and supported by nature" [59]. The European Commission [60] has identified four principal goals for nature-based solution: (1) enhance sustainable urbanization to stimulate economic growth as well as improve the environment, make cities more attractive, and enhance human well-being; (2) restore degraded ecosystems to improve the resilience of ecosystems, enable them to deliver vital ecosystem services, and also meet other societal challenges; (3) develop climate change adaptation and mitigation to provide more resilient responses; and (4) improve risk management and resilience which can lead to greater benefits that supersede conventional methods and offer synergies in reducing multiple risks.

An urban context example of such a solution is the Semiahmoo Library in Surrey, British Columbia, Canada which houses a living wall of over 10,000 individual plants and more than 120 species (Fig. 10.3). The library's exterior is a self-sufficient green wall with veritable shrubbery sustained by water and nutrients delivered directly

Fig. 10.3 Semiahmoo Library living wall located in Surrey, British Columbia, Canada. Photographs taken by G. T. Cirella on September 20, 2018

from within its vertical support system. This living wall, also called a bio-wall, is composed of pre-vegetated panels, vertical modules, and planted blankets [61]. The wall is constructed of three parts: metal frame, polyvinyl chloride layer, and air layer (i.e., there is no soil layer). This type of wall would support a wide variety of environments and climatic conditions. In addition, "living" architecture as in Semiahmoo Library normally has a number of secondary functions such as noise reduction (i.e., the ability to suppressive sound), insulative properties, and habitat and relating services for insects (e.g., bees) and animals (e.g., birds). Nature-based solutions are, to some degree, an all-encompassing concept. They integrate a system understanding (i.e., ecology and society) and values and benefits (i.e., biodiversity, resource efficiency, food security, disaster risk reduction, clean water, health and well-being, clean air, and climate control) with ecosystem functions and services (e.g., food, raw material, waste decomposition, recreation, climate regulation, water storage and filtration, disease control, coastal protection, and aesthetics). Noticeably, this array of interdisciplinarity places nature-based solutions at an important point of

Table 10.1 Urban ecological corridors and related concepts, adapted from Peng et al. [62]

Concept	Characteristic
Greenway	Linear landscape owning multiple functions such as ecological, cultural, recreational, and aesthetic functions
Green belt	Green open space set up in urban peripheral, used for urban and rural segmentation
Habitat and ecological network	Linear or strip landscape with the ecological, social, cultural, and other functions
Ecological infrastructure	Reticular landscape or open space with basic ecosystem services consisting of point, line, and surface

the design phase within urban green infrastructure and how subsequent future urban trends, if carefully implemented, can greatly benefit the wider urban and peri-urban ecology of cities.

2.4 Urban Ecological Corridors

The concepts related to urban ecological corridors include greenways, green belts, green corridors, ecological networks, habitat networks, and ecological infrastructure [62] (Table 1). There are also diverse classifications of urban ecological corridors (e.g., river corridors, green transportation corridors, biodiversity conservation corridors, heritage corridors, and recreation corridors) in which complexity in the structure and function [62] has linearity and linkage in common [63]. In cities, corridors have an important function in the conservation of biodiversity in urban domestic gardens and effectiveness in terms of the distribution of organisms with low dispersal capabilities [64]. Planning an urban ecological network can reduce landscape fragmentation and increase the shape complexity of green space patches and landscape connectivity [65]. The health benefits of green corridors have been demonstrated to be psychologically as well as biophysically beneficial [63]. Urban planners and landscape architects need ecological information to plan and design for location, spacing, and dimensions of corridors [66]. Austin [66] notes that a 10–30 m wide corridor could be sufficient for species moderately or highly adapted to human activity; however, these minimum dimensions may require site-specific adjustments (e.g., variability of the animal species, climate, vegetation, and topography).

2.5 Urban Biosphere Reserve

The International Coordinating Council of the Man and the Biosphere (MAB) Programme, a UNESCO working group, identified four development themes for urban biosphere reserves: zonation of biosphere reserves; governance of biosphere reserves;

policy, management, and business plans; and data management and monitoring [67]. Russo and Cirella [50] extrapolated MAB's research into different urban biosphere reserve typologies in which four principle urban settings are identified (Fig. 10.4). According to Dogse [68], urban biosphere reserves are "important urban areas within or adjacent to its boundaries where the natural, socioeconomic, and cultural environments are shaped by urban influences and pressures and set up and managed to mitigate these pressures for improved urban and regional sustainability."

An example of an urban biosphere reserve is the Dublin Bay Biosphere Reserve in Ireland which includes North Bull Island, Dublin Bay, and the adjacent land which includes parts of Dublin. This enlarged biosphere is made up of grassland dunes, sand dunes and salt marsh with glasswort (*Salicornia dolichostachya* and *S. europaea*), *Puccinellia maritima*, and sea lavender (*Limonium humile*) (Fig. 10.5). It is one of the most pristine sand dunes in Ireland and globally, a first-rate example of an urban region biosphere reserve [69].

Another two noteworthy examples of urban biosphere reserves come emanate from England and Australia. First, the Brighton and Lewes Downs Biosphere Reserve located on the Southeast coast of England is apart a central unit of the hills of the South

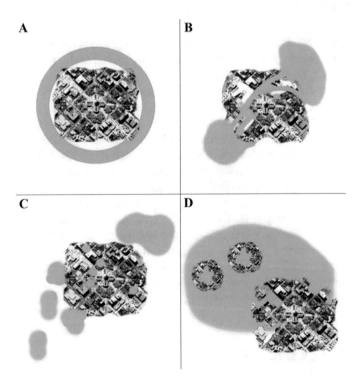

Fig. 10.4 Urban biosphere reserve typologies where **a** urban green belt, **b** urban green corridor, **c** urban green cluster, **d** urban region. Adopted from MAB [67] and reproduced from Russo and Cirella [50] using Paint.Net 4.24

Fig. 10.5 (left) Dublin Bay biosphere reserve grassland dunes where important orchids such as the (middle) bee orchid and (right) pyramidal orchid grow. Photographs taken by A. Russo on June 21, 2019

Downs National Park [70]. The reserve is centered on the Brighton chalk block that is situated between the River Adur and River Ouse. Chalk downland forms much of the terrestrial landscape, while the coastline is dominated by chalk cliffs and urbanized plains [70]. Second, the Great Sandy Biosphere located along the Fraser Coast region of Queensland, Australia, encompasses "the neighboring Gympie area and the Bundaberg coastline and puts it in the same class as the Galapagos Islands, the Central Amazon, the Everglades, and Uluru" [71]. The Great Sandy Biosphere protects natural resources while offering a balanced conservation and sustainable development platform. Nature conservation (e.g., major breeding for endangered marine turtles, the largest sand mass in the world, and some 7500 species of fauna and flora) is coupled in with the cities of Hervey Bay, Maryborough, and Gympie (Fig. 10.6).

Fig. 10.6 Hervey Bay's inlet living, a part of the Great Sandy Biosphere reserve along the Fraser Coast region of Queensland, Australia. Photograph taken by G. T. Cirella on May 18, 2014

3 Case Examples

3.1 New York City: "Green Infrastructure Plan"

The Green Infrastructure Plan presents an alternative approach to improving water quality by integrating green infrastructure. The Plan incorporates swales and green roofs, with investments to optimize the existing water system to build targeted, smaller-scale gray, or traditional infrastructure. This is a multi-pronged, modular, and adaptive approach to a complicated problem that can provide widespread, immediate benefits at a lower cost. The green infrastructure component of this strategy builds upon and reinforces strong public and government support necessary to make additional water quality investments.

One critical goal of the Green Infrastructure Plan is to manage runoff from 10% of the impervious surfaces in combined sewer watersheds through detention and infiltration source controls [72]. This blue print builds upon and extends the commitment of the sustainable stormwater management plan framed between New York Harbor and New York City. The Plan's strategy and operating costs are projected help with "reduction strategies and investments over the next 20 years and will lead to both clean waterways and a greener, more sustainable city" [72]. Similarly, to New York City, other financial capitals like London, Hong Kong, Singapore, and Shanghai have been developing their own green urban strategies that show promise and imagination in integrating ecology in city design and planning.

3.2 London: "City Plan 2036"

The City of London's Planning and Transportation Committee recently backed an ambitious local plan called "City Plan 2036." This plan's intention protects existing open spaces and requires development to provide new spaces, where feasible. The advancement and change oblige development to focus on green walls and roofs, trees, and other green features (Fig. 10.7). The Plan's objective is to assess the level of greening through a new "Urban Greening Factor which will be applied to major new development" [73]. As such, building designs and the public realm is to focalize on long-term resilience infrastructure that can withstand a variety of climate conditions (e.g., overheating and flooding). To facilitate this move, all development, transport, and public infrastructural designs must integrate with the city's sustainable urban drainage principles [73]. In London, there a common concern that the city is overly dependent on the financial and business services sector, so much so, that such that City Plan 2036 has refocused its stratagem on "diversity and resilience, promoting strong performance across more of the economy, with no single sector contributing more than 40% of gross value added or jobs growth" [74]. London's green infrastructure strategies are future-oriented and exploratory for large-scale, urbanized areas that plan to develop or implement urban-based sustainability with green-friendly design.

Fig. 10.7 London Wall Place. Photograph taken by A. Russo on June 15, 2018

4 Conclusion: Sustainable Planning and Design Tools for Measuring, Modeling, and Valuing Ecosystem Services

Many previous studies have shown that urbanization can impact not only the spatial pattern of urban green infrastructure but also the urban landscape itself which, in turn, influences its ecosystem services [75]. To better analyze and evaluate urban green infrastructure, its spatial patterns and influence on urban sustainability, we need appropriate approaches [75]. In the last few years, ecosystem services of urban green infrastructure have been assessed using available methods including computer-based modeling tools—software or web-based tools such as ENVI-met, i-Tree (i.e., Tree Eco, i-Tree Canopy, and i-Tree Landscape), and in the past, CITY green [44, 76]. These models have been developed or adapted for landscape design and urban planning [44]. There are an increasing number of tools available that aim to value urban green infrastructure such as the non-departmental public body in the United Kingdom's Natural England [77] which assessed a number of these tools in terms of their adherence to the principles of scientific and economic analysis and their applicability within to their country.

Several of these tools and applications use tree inventory data to quantify the monetary and non-monetary value of environmental and aesthetic benefits generated from urban green infrastructure [78]. Often, these available models account for ecosystem functions that are detrimental to human well-being and the related costs of management, thus allowing for a better understanding of the urban forest and strategic planning, but regrettably, they do not consider any specific ecosystem disservices to balance against the ecosystem services [78]. Scientists, city managers, and policy-makers alike need to be aware that available models and methods can produce statistically different ecosystem service estimates [79]. Evaluation and monitoring of urban green infrastructure need to use the appropriate level of scientific rigor [52] in combination with ecologically oriented thinking.

Investing in urban green infrastructure in cities, including ecological restoration and rehabilitation of ecosystems (e.g., urban rivers, lakes, and woodlands), is not only ecologically and socially desirable but also economically advantageous [80]. Urban sustainability integrates a number of overlapping concepts and methods. To evaluate ecosystem services for sustainable urban design and planning and to further advance our understanding of the complex socio-ecological processes, combined with rising levels of urbanization, we need to integrate primary research into urban design processes [81]. Best systems control and practice should support sustainable ecological planning in conjunction with urban strategies and green scholarship.

References

1. Morse NB, Pellissier PA, Cianciola EN et al (2014) Novel ecosystems in the Anthropocene: a revision of the novel ecosystem concept for pragmatic applications. Ecol Soc 19:art12. https://doi.org/10.5751/ES-06192-190212
2. Lu X, Lin C, Li W et al (2019) Analysis of the adverse health effects of PM2.5 from 2001 to 2017 in China and the role of urbanization in aggravating the health burden. Sci Total Environ 652:683–695. https://doi.org/10.1016/j.scitotenv.2018.10.140
3. Aronson MFJ, Sorte FA La, Nilon CH et al (2014) A global analysis of the impacts of urbanization on bird and plant diversity reveals key anthropogenic drivers
4. Concepción ED, Moretti M, Altermatt F et al (2015) Impacts of urbanisation on biodiversity: the role of species mobility, degree of specialisation and spatial scale. Oikos 124:1571–1582. https://doi.org/10.1111/oik.02166
5. Liu Z, He C, Wu J (2016) The relationship between habitat loss and fragmentation during urbanization: an empirical evaluation from 16 world cities. PLoS ONE 11:e0154613. https://doi.org/10.1371/journal.pone.0154613
6. Yigitcanlar T, Dizdaroglu D (2015) Ecological approaches in planning for sustainable cities. A review of the literature. Glob J Environ Sci Manag 1:159–188. https://doi.org/10.7508/gjesm.2015.02.008
7. Wang X, Palazzo D, Carper M (2016) Ecological wisdom as an emerging field of scholarly inquiry in urban planning and design. Landsc Urban Plan 155:100–107. https://doi.org/10.1016/j.landurbplan.2016.05.019
8. Wu J (2014) Urban ecology and sustainability: the state-of-the-science and future directions. Landsc Urban Plan 125:209–221. https://doi.org/10.1016/j.landurbplan.2014.01.018
9. Childers DL, Pickett STA, Grove JM et al (2014) Advancing urban sustainability theory and action: challenges and opportunities. Landsc Urban Plan 125:320–328. https://doi.org/10.1016/j.landurbplan.2014.01.022
10. Ostrom E (1990) Governing the commons: The evolution of institutions for collective action. Cambridge University Press
11. Patten DT (2016) The role of ecological wisdom in managing for sustainable interdependent urban and natural ecosystems. Landsc Urban Plan 155:3–10. https://doi.org/10.1016/j.landurbplan.2016.01.013
12. Cirella GT, Zerbe S (2014) Index of sustainable functionality: procedural developments and application in Urat front banner, Inner Mongolia autonomous region. Int J Environ Sustain 10:15–31

13. Steiner F (2014) Frontiers in urban ecological design and planning research. Landsc Urban Plan 125:304–311. https://doi.org/10.1016/j.landurbplan.2014.01.023

14. Bogar S, Beyer KM (2016) Green space, violence, and crime. Trauma, Violence, Abus 17:160–171. https://doi.org/10.1177/1524838015576412

15. Troy A, Morgan Grove J, O'Neil-Dunne J (2012) The relationship between tree canopy and crime rates across an urban–rural gradient in the greater Baltimore region. Landsc Urban Plan 106:262–270. https://doi.org/10.1016/j.landurbplan.2012.03.010

16. Escobedo FJ, Clerici N, Staudhammer CL et al (2018) Trees and crime in Bogota, Colombia: is the link an ecosystem disservice or service? Land Use Policy 78:583–592. https://doi.org/10.1016/j.landusepol.2018.07.029

17. Eisenman TS (2013) Frederick Law Olmsted, green infrastructure, and the evolving city. J Plan Hist 12:287–311. https://doi.org/10.1177/1538513212474227

18. Howard E (1898) Garden cities of tomorrow (being the second edition to tomorrow: a peaceful path to real reform). Swan Sonnenschein and Co., Ltd., London

19. George H (1881) Progress and poverty: an inquiry into the cause of industrial depressions and of increase of want with increase of wealth. The Remedy. D. Appleton and Company, New York

20. Abercrombie P, Nickson R (1949) Warwick: its preservation and redevelopment. Architectural Press, Warwick Borough Council

21. Jacobs J (1961) The death and life of great American cities. Modern Library, New York

22. Nadal A, Cerón I, Cuerva E et al (2015) Urban agriculture in the framework of sustainable urbanism. Elisava Temes de disseny 0:92–103

23. Habitat UN (2009) Planning sustainable cities: global report on human settlements. Earthscan, London

24. van Dijk MP (2011) Three ecological cities, examples of different approaches in Asia and Europe. In: Wong T-C, Yuen B (eds) Eco-city Planning. Springer, Netherlands, Dordrecht, pp 31–50

25. Joss S (2015) Eco-cities and sustainable urbanism. Int Encycl Soc Behav Sci 6:829–837. https://doi.org/10.1016/B978-0-08-097086-8.74010-4

26. City of Helsinki (2005) Eco-Viikki, aims implementation and results. City of Helsinki, Vantaa

27. Marcus L (2018) Overcoming the subject-object dichotomy in urban modeling: axial maps as geometric representations of affordances in the built environment. Front Psychol 9:1–10. https://doi.org/10.3389/fpsyg.2018.00449

28. Young RF, Lieberknecht K (2018) From smart cities to wise cities: ecological wisdom as a basis for sustainable urban development. J Environ Plan Manag 0:1–18. https://doi.org/10.1080/09640568.2018.1484343

29. Whitehead M (2003) (Re)analysing the sustainable city: nature, urbanisation and the regulation of socio-environmental relations in the UK. Urban Stud 40:1183–1206. https://doi.org/10.1080/0042098032000084550

30. Zhang X, Li H (2018) Urban resilience and urban sustainability: what we know and what do not know? Cities 72:141–148. https://doi.org/10.1016/j.cities.2017.08.009

31. Chelleri L, Schuetze T, Salvati L (2015) Integrating resilience with urban sustainability in neglected neighborhoods: challenges and opportunities of transitioning to decentralized water management in Mexico City. Habitat Int 48:122–130. https://doi.org/10.1016/j.habitatint.2015.03.016

32. Fitzgibbons J, Mitchell CL (2019) Just urban futures? Exploring equity in "100 resilient cities". World Dev 122:648–659. https://doi.org/10.1016/j.worlddev.2019.06.021

33. UNEP (1972) Declaration of the United Nations conference on the human environment. United Nations Environmental Programme, Stockholm

34. United Nations (2017) The sustainable development goals report 2017, New York

35. Bibri SE (2019) On the sustainability of smart and smarter cities in the era of big data: an interdisciplinary and transdisciplinary literature review. J Big Data 6:25. https://doi.org/10.1186/s40537-019-0182-7

36. World Bank (2012) Who needs smart cities for sustainable development? In: Feature story. https://www.worldbank.org/en/news/feature/2012/03/20/who-needs-smart-cities-for-sustainable-development. Accessed 31 Jul 2019
37. Basiri M, Azim AZ, Farrokhi M (2017) Smart city solution for sustainable urban development. Eur J Sustain Dev 6:71–84. https://doi.org/10.14207/ejsd.2017.v6n1p71
38. Ahvenniemi H, Huovila A, Pinto-Seppä I, Airaksinen M (2017) What are the differences between sustainable and smart cities? Cities 60:234–245. https://doi.org/10.1016/j.cities.2016.09.009
39. Haarstad H, Wathne MW (2019) Are smart city projects catalyzing urban energy sustainability? Energy Policy 129:918–925. https://doi.org/10.1016/j.enpol.2019.03.001
40. Zhu S, Li D, Feng H (2019) Is smart city resilient? Evidence from China. Sustain Cities Soc 50:101636. https://doi.org/10.1016/j.scs.2019.101636
41. UNESCO (2014) United Nations economic and social council integration segment. In: United Nations. https://www.un.org/ecosoc/en/ecosoc-integration-segment. Accessed 31 Jul 2019
42. Kaur H, Garg P (2019) Urban sustainability assessment tools: a review. J Clean Prod 210:146–158. https://doi.org/10.1016/j.jclepro.2018.11.009
43. EEA (2010) The European environment: state and outlook 2010: Urban Environment, Luxembourg
44. Russo A, Escobedo FJ, Zerbe S (2016) Quantifying the local-scale ecosystem services provided by urban treed streetscapes in Bolzano, Italy. AIMS Environ Sci 3:58–76. https://doi.org/10.3934/environsci.2016.1.58
45. Brown RD, Gillespie TJ (1995) Microclimatic landscape design: creating thermal comfort and energy efficiency, 1st edn. Wiley
46. Chen D, Thatcher M, Wang X et al (2015) Summer cooling potential of urban vegetation—a modeling study for Melbourne, Australia. AIMS Environ Sci 2:648–667. https://doi.org/10.3934/environsci.2015.3.648
47. Ng E, Chen L, Wang Y, Yuan C (2012) A study on the cooling effects of greening in a high-density city: an experience from Hong Kong. Build Environ 47:256–271. https://doi.org/10.1016/j.buildenv.2011.07.014
48. Peng L, Jim C (2013) Green-roof effects on neighborhood microclimate and human thermal sensation. Energies 6:598–618. https://doi.org/10.3390/en6020598
49. Hofmann M, Young C, Binz TM et al (2017) Contact to nature benefits health: mixed effectiveness of different mechanisms. Int J Environ Res Public Health 15. https://doi.org/10.3390/ijerph15010031
50. Russo A, Cirella G (2018) Modern compact cities: how much greenery do we need? Int J Environ Res Public Health 15:2180. https://doi.org/10.3390/ijerph15102180
51. Russo A, Cirella GT (2018) Edible green infrastructure 4.0 for food security and well-being: Campania Region, Italy. In: Quinlan V (ed) International guidelines on urban and territorial planning. Compendium of Inspiring Practices: Health Edition. UN Habitat, HS/080/18E, Nairobi, Kenya, p 72
52. European Commission's Directorate-General Environment (2012) The multifunctionality of green infrastructure
53. Franchini M, Mannucci PM (2018) Mitigation of air pollution by greenness: a narrative review. Eur J Intern Med 55:1–5. https://doi.org/10.1016/j.ejim.2018.06.021
54. Murthy VH (2016) Food insecurity: a public health issue. Public Health Rep 131:655–657. https://doi.org/10.1177/0033354916664154
55. Russo A, Escobedo FJ, Cirella GT, Zerbe S (2017) Edible green infrastructure: an approach and review of provisioning ecosystem services and disservices in urban environments. Agric Ecosyst Environ 242:53–66. https://doi.org/10.1016/j.agee.2017.03.026
56. Russo A, Cirella GT (2018) Edible green infrastructure for urban regeneration: case studies from the Campania Region, Italy. In: 4th international symposium on infrastructure development, 12–14 Oct 2018. Hasanuddin University and Manado State Polytechnic, Manado, Indonesia
57. Nesshöver C, Assmuth T, Irvine KN et al (2017) The science, policy and practice of nature-based solutions: an interdisciplinary perspective. Sci Total Environ 579:1215–1227. https://doi.org/10.1016/j.scitotenv.2016.11.106

58. Lafortezza R, Chen J, van den Bosch CK, Randrup TB (2018) Nature-based solutions for resilient landscapes and cities. Environ Res 165:431–441. https://doi.org/10.1016/j.envres.2017.11.038
59. Raymond CM, Frantzeskaki N, Kabisch N et al (2017) A framework for assessing and implementing the co-benefits of nature-based solutions in urban areas. Environ Sci Policy 77:15–24. https://doi.org/10.1016/j.envsci.2017.07.008
60. European Commission (2015) Towards an EU research and innovation policy agenda for nature-based solutions & re-naturing cities
61. Timur ÖB, Karaca E (2013) Vertical gardens. In: Ozyavuz M (ed) Advances in landscape architecture. InTech, London, pp 587–622
62. Peng J, Zhao H, Liu Y (2017) Urban ecological corridors construction: a review. Acta Ecol Sin 37:23–30. https://doi.org/10.1016/j.chnaes.2016.12.002
63. Ignatieva M, Stewart GH, Meurk C (2011) Planning and design of ecological networks in urban areas. Landsc Ecol Eng 7:17–25. https://doi.org/10.1007/s11355-010-0143-y
64. Vergnes A, Kerbiriou C, Clergeau P (2013) Ecological corridors also operate in an urban matrix: a test case with garden shrews. Urban Ecosyst 16:511–525. https://doi.org/10.1007/s11252-013-0289-0
65. Li H, Chen W, He W (2015) Planning of green space ecological network in urban areas: an example of Nanchang, China. Int J Environ Res Public Health 12:12889–12904. https://doi.org/10.3390/ijerph121012889
66. Austin GD (2012) Multi-functional ecological corridors in urban development. Spaces Flows An Int J Urban ExtraUrban Stud 2:211–228. https://doi.org/10.18848/2154-8676/CGP/v02i03/53662
67. UNESCO (2018) MAB programme. In: UNESCO MAB, Man Biosph. Program. http://www.unesco.org/new/en/natural-sciences/environment/ecological-sciences/. Accessed 22 Feb 2018
68. Dogse P (2004) Toward urban biosphere reserves. Ann N Y Acad Sci 1023:10–48. https://doi.org/10.1196/annals.1319.002
69. UNESCO (2015) Dublin bay. In: UNESCO MAB, Man Biosph. Program. http://www.unesco.org/new/en/natural-sciences/environment/ecological-sciences/biosphere-reserves/europe-north-america/ireland/dublin-bay/. Accessed 25 Jul 2019
70. UNESCO (2014) Brighton and Lewes Downs. New York
71. Queensland Government (2019) Great sandy biosphere. In: Tour. Events Queensl. https://www.queensland.com/en-gb/attraction/great-sandy-biosphere. Accessed 18 May 2019
72. New York City (2010) NYC green infrastructure plan. New York City Environmental Protection, New York
73. City of London (2018) City plan 2036. City of London, London
74. Ferm J, Jones E (2017) Beyond the post-industrial city: valuing and planning for industry in London. Urban Stud 54:3380–3398. https://doi.org/10.1177/0042098016668778
75. Breuste J, Artmann M, Li J, Xie M (2015) Special issue on green infrastructure for urban sustainability. J Urban Plan Dev 141:A2015001. https://doi.org/10.1061/(ASCE)UP.1943-5444.0000291
76. Neugarten RA, Langhammer PF, Osipova E et al (2018) Tools for measuring, modelling, and valuing ecosystem services: guidance for Key Biodiversity Areas, natural World Heritage sites, and protected areas. IUCN, Gland, Switzerland
77. England Natural (2013) Green infrastructure: valuation tools assessment. Natural England, Exeter
78. Speak A, Escobedo FJ, Russo A, Zerbe S (2018) An ecosystem service-disservice ratio: using composite indicators to assess the net benefits of urban trees. Ecol Indic 95:544–553. https://doi.org/10.1016/j.ecolind.2018.07.048
79. Russo A, Escobedo FJ, Timilsina N et al (2014) Assessing urban tree carbon storage and sequestration in Bolzano, Italy. Int J Biodivers Sci Ecosyst Serv Manag 10:54–70. https://doi.org/10.1080/21513732.2013.873822

80. Elmqvist T, Setälä H, Handel SN et al (2015) Benefits of restoring ecosystem services in urban areas. Curr Opin Environ Sustain 14:101–108. https://doi.org/10.1016/j.cosust.2015.05.001
81. Felson AJ, Pavao-Zuckerman M, Carter T et al (2013) Mapping the design process for urban ecology researchers. Bioscience 63:854–865. https://doi.org/10.1525/bio.2013.63.11.4

Urbanization and Population Change: Banjar Municipality

Agus Supriyadi, Tao Wang, Shanshan Chu, Tianwu Ma,
Raden G. Shaumirahman, and Giuseppe T. Cirella

Abstract There is a worldwide growing concern that increased concentration of people are moving to cities. There are numerous studies that describe and explain this phenomenon. A limited amount of research focuses on specific autonomous urban environments. Evidence on land use and population density change examines Banjar Municipality, a new autonomous city, in Indonesia. Google Earth Image Series data, an appropriate solution for tracking land use change in smaller geographic areas, is examined between 2006 and 2016. Results show an increase in urban area from 13.49 to 15.41% while agricultural land decreased from 71.22 to 69.87%. Positive observational highlights indicate urban forested areas are minimally impacted by the Municipality's city-wide development. A small percentage of the surrounding water bodies increased as a result of the local authority building a new artificial lake. Interestingly, urban area changes mostly have occurred throughout the districts of Banjar and Langensari, not in Pataruman, which has experienced the highest increase in population in the latter decade. The underlying research focalizes on the constraints to facilitate land use within an urban corridor setting.

A. Supriyadi (✉) · T. Wang · S. Chu · T. Ma
School of Geography, Nanjing Normal University, Nanjing, China
e-mail: agus_dea@yahoo.com

T. Wang
e-mail: wangtao@njnu.edu.cn

S. Chu
e-mail: 995257156@qq.com

T. Ma
e-mail: tianwuma@foxmail.com

R. G. Shaumirahman
Urban and Regional Engineering Faculty, Pasundan University, Bandung, Indonesia
e-mail: guswinden@gmail.com

G. T. Cirella
Faculty of Economics, University of Gdansk, Sopot, Poland
e-mail: gt.cirella@ug.edu.pl

© Springer Nature Singapore Pte Ltd. 2020
G. T. Cirella (ed.), *Sustainable Human–Nature Relations*,
Advances in 21st Century Human Settlements,
https://doi.org/10.1007/978-981-15-3049-4_11

Keywords Urban corridor · Population density change · Land use · Autonomous city

1 Introduction

Rapid changes in the development have influenced the paradigm of development systems in every country. According to the 2016 Habitat III Conference on cities, it emphasized worldwide urban populations are predicted to increase significantly in the middle of this century. As such, urban problems (e.g., socioeconomic and cultural concerns) will be concentrated in cities, effecting the environment and yielding potential for humanitarian crises. Huilei et al. [1] highlight the challenges of a future where four out of every five people live in peri-urban and urban environments. Key hardships include housing, infrastructure, and food security. Asia encompasses the largest landmass, populace, and built-up areas of urbanization worldwide with 57.7%, followed by North America (i.e., 12.4%), Africa (i.e., 11.4%), and Europe (i.e., 9.8%), respectively [2]. After the Asian financial crisis hampered much of the continent, by 1997 a massive transformation began to take place in many of its most populous countries (i.e., China, India, Thailand, and Indonesia). Tangible evidence from Indonesia can be observed for the decade following the year 2000, in which Indonesian cities grew faster compared to any other Asian country. This transformation, from rural (i.e., mostly agriculture-based) to urban, forced a rethinking from a societal, developmental, and demographic standpoint. According to the World Bank, Indonesia's population increased an average of 4.1% compared to 3.8% in China, 3.1% in India, and 2.8% in Thailand [3]. Interestingly, during this first decade of the new millennium, Indonesia experienced a growth in urban land added per new urban resident at less than 40 m^2, which is considered the smallest amount of any country Asia-wide [3]. As a result, it is undeniable Indonesia encountered and continues to deal with the large facet of challenges directly connected with these changes.

Throughout Indonesia, a number of small- and medium-sized cities have emerged near major cities creating a dependency mechanism on the major city. This type of city development is not exclusively urbanization, including other factors such as comparative advantages (i.e., topography, natural resources, and historical connections) and construction development (i.e., infrastructure network and social facilities) [4]. Moreover, the pattern of spatial development in urban populations shows evidence of a new phenomenon of urban development throughout the country: urban corridors. Particularly on the island of Java (i.e., where the country's highest concentration of people lives), there has been a tendency to develop connective corridors between large urban centers, including: Serang–Jakarta–Karawang corridor, Jakarta–Bandung corridor, Cirebon–Semarang corridor, Semarang–Yogyakarta corridor, and Surabaya–Malang corridor. The formation of these urban corridors clearly alters the urban and, particularly, rural areas that lie between connective points [4]. Ardiwijaya et al. [5] explained sub-urbanization as random sprawl that continues to grow uncontrollably. Sub-urbanization further blurs and makes it difficult to distinguish boundaries between urban and rural areas.

Over a period of time, socioeconomic change and innovation can encourage the transition of land use [6]. It can be observed in terms of quantity, structure, and spatial patterns which are similar to landscape structures (i.e., examples of dominant land use morphology) [7]. Evidence from China shows that through trial and error, the development of rural areas critical in achieving successful rural construction and land consolidation. Local inhabitants are given the opportunity to take part in decision making for future development [8]. Based on the evidence, we can see that empowering local authorities and inhabitants is very important in achieving better opportunities directly significant to the success of development [9, 10]. This is evident within areas dominated by human-built environment, where it combines all non-vegetative elements, including: roads, buildings, runways, and industrial facilities [11]. In an urban analysis, there is often a discrepancy between planned development by municipal governments and actual growth occurring in the peri-urban areas of cities [12]. This condition usually occurs when the development of the city is taken over for a political purpose—especially when national or local leadership changes. Urban land use change is one of the vital keys to understanding the success of sustainable development planning and to evaluate how urban areas grow, especially in correlation to socioeconomic disparity.

The gap between urban and rural areas has become one of the key factors that influence migration from undeveloped areas to more developed ones [13]. This situation can be interlinked with the expectation to stimulate economic development in rural areas by empowering local authorities to develop the local economy. Many studies conducted in several countries have observed the impact of urbanization on environment and land use impact in big cities; however, insufficient research has been deduced when considering autonomous cities which are closely connected with rural environments. In fact, there is an additional gap in the literature when considering urban area change of cities from non-autonomous to autonomous.

In terms of land use imagery, limitations in attaining free, high resolution, and access-friendly (i.e., most up-to-date) remote sensing images are evident in many studies—often blocked due to security and government regulation. Google Earth imagery is one reasonable solution that is open source, providing clear imagery (e.g., buildings and roads) for a number of urban-related applications [14–16]. The use of remote sensing and geographic information systems (GIS) analysis in urban-related research has been carried out in Indonesia since early 2000. As such, these techniques in the last four years have focalized on understanding the environmental and socioeconomic conditions of Semarang in Central Java, including modeling of it urban development [17] and urban growth in relation to population [18] and land use [12]. Still, very limited research has been conducted when considering the autonomous city setting. In response to this gap, Banjar Municipality, an autonomous city (i.e., not located in any regency) along the eastern border of West Java Province, is the focus of this chapter. We utilize imagery software from 2018 and mapping data of Banjar Municipality urban corridor to illustrate changes to its land use, urban area, and population.

2 Autonomous Cities in Indonesia

2.1 *Response to Globalization*

Globalization is a series of unifying processes of structuration and stratification, multifaceted social phenomena of removing political boundaries or deterritorialization and reterritorialization of socioeconomic and political space. This influence is undeniably encroaching upon all aspects of society—worldwide. The age of technological advancement continues to interlink information at unprecedented levels [19]. For Indonesia, positive and negative impacts can be observed. Positive effects include infrastructure and economic development which have not drastically altered traditional Indonesian culture. Adverse effects, however, include the upsurge damaging occurrence of war and internal conflict, especially as a result of the transition from older forms of globalization to the new [20]. Economic rivalry and threat of national unity are two fundamental characteristics of new era globalization [21]. In Indonesia, this setting has become more prevalent as the world and international markets continue to integrate and connect in ways in which virtually no physical boundaries coexist. Even so, key impacts from globalization in Indonesia spurred from the 1990s with the introduction of trade liberalization region-wide—significantly decreasing child labor and related forms of exploitation [22].

With the development of technology and increase of social and economic interaction between the individual and organization, new era globalization trumps the state and its physical control of force [23]. As a result of low-cost communication and its transformation to enable population mobility and cross-border communication, the world has become more integrated by interactions via environmental, social, political, and, to a larger part, economic means. Global flows of commodities, information, capital, and people, as well as factors from remote markets, are increasingly the impetus of land use change. This condition is often associated with a growing consumer class in developing countries. By 2005, when Indonesia was implementing its new reformation era, it was ranked ninth in the new globalization index with the score of 18,02 for Asian countries followed by Japan with the score 17,71 [24]. It is undeniable a high-level connection exists between globalization and other socioeconomic phenomena such as poverty, economic growth, infrastructural development, inequality, and living standard. Globalization no longer discussed at the national level where inter-state actors nearly have no partition has moved to lower regional levels (i.e., province, district, and city). It can be further be deduced that regional autonomy is also closely linked to globalization that emerged in the 1970s and 1980s in the USA. Since decentralization and regional autonomy was implemented in Indonesia, it has faced a number of obstacles in which development of any new autonomous region would involve a vision of borderless regional planning [25]. Globalization does not yet have an established definition of success since it deeply depends on a society's perspective and level of embracement. As such, evidence shows that an autonomous city tends to have a higher population growth rate compared to a non-autonomous one; thus, improving the quality of government becomes very important for economic development [26] via increased levels of self-government. Regional development and decentralized institutions encourage local public authorities to become more active in developing their environment with local-based knowledge having a positive and measurable impact on improving governance [27]. In the governance of decentralization, a new paradigm of the political governance success is not necessarily classic development such as roads, schools, or any other infrastructure, but measured by how projects enhance the reputation of political leaders who fund them [28].

The increase of regionalism in the framework of increasing economic growth and improving the overall pattern of globalization gave birth to the policy of regional autonomy in Indonesia. Interesting pieces of evidence that support this idea is the impact policy had on the regional autonomy in Karawang Regency, West Java Province. Karawang is famous for its national rice granary; however, since its status as regional autonomy was implemented, the city has transformed itself into a manufacturing-base, suppressing the original culture of indigenous life (i.e., farming and fishing) with factories and industrial work [25]. Moreover, the land use massively altered the Karawang rice paddy fields from 21,909 ha in 2000 to 16,355 ha 2015 [29].

2.2 Decentralization and Regional Autonomy

Historically, before Indonesia declared independence in 1945, the practice of decentralization had been attempted with various motivations, including colonialists seeking to maintain colonial trade by restricting control from local leaders. The first Indonesian president tried to balance this structure [30]. In practice, decentralization since independence has been mostly centralized by the central government, in which all policies are controlled and regulated. The provincial and local authorities, as an extension, have not had the authority to regulate their territory and cannot carry out regional development initiatives. In 1974, Law No. 5, concerning regional government (i.e., *pemerintah daerah*), was born which regulates the relationship between central and local governments. It was formally based on the principle of decentralization, deconcentration, and co-administration, but again in practice not fully implemented or applied [30]. As such, the structure of provincial and local governments in Indonesia has been overridden by national-level politics mostly focused on integration and stability.

In 1997, when the financial crisis hampered Asia, Indonesia experienced enormous impact leading to the resignation of its second president, President Suharto. Since then, there have been massive national political changes, including the system of governance which is commonly known as the reformation period in Indonesia. The reformation influenced all aspects of governance. One significant change was the implementation of decentralization. One year in, in 1998, disintegrating conditions and pressure from international donor countries as well as national parliament drafted decentralizing laws in hope of more democratic ideals. As a result, the new regulation No. 22/1999 on regional government was enacted, followed by Act No.

25 on fiscal balance between central and local government. The legalized standard effectively was implemented in January 2001. Throughout the country, its territory was divided into autonomous provinces, districts or regencies (i.e., *kabupaten*), and municipalities (i.e., *kota*). Districts and municipalities are technically the same level of government. This distinction is based on whether the government administration is located in a rural area (i.e., regency) or an urban area (i.e., municipality). Within districts and municipalities, there are districts (i.e., *kecamatan*) which are smaller administrative government units. Each district is further divided into villages. Villages in rural areas are called *desa*, while in urban areas they are referred to as *kelurahan*. From 1985 to 2015, over a period of 30 years, newly implemented autonomous regions, provinces, districts, cities, and villages have almost doubled (Table 1).

The practice for decentralized policies to achieve optimal results requires a significant time for maturation [31, 32]. Its application should be calculated with a considerable capacity to increase individual, institutional, and system-oriented levels. Implementation of the value of the decentralization in Indonesia is considered less successful since the data indicates (i.e., about 80% of new autonomous regions are considered untouched from the outside) [33]. Since the regional autonomy policy was implemented, autonomous urban areas Indonesia-wide have been struggling to meet opportunity and restrictive norms simultaneously [34, 35]. This condition, instead, has sparked regional divisions of authority interrupting city growth and development with that more in line with the needs of the region. However, this condition has augmented competition between regions and the public service system has become more inclined with ego-regionality [36, 37].

Empiric evidence can be seen in small towns throughout Central Java where concentric structured units have slow land development due to limited economic stimulus (i.e., usually restricted to a single activity). On the other hand, multiple nuclei structured small towns can be quite developed, ranging from moderate to high standards of living. This situation is due to the existence of several centers result work together to development and built-up land based on spatial typology patterns of coexistence. This situation is evident throughout small towns located on the outskirts of major cities such as Surakarta, Yogyakarta, Semarang, Tegal, Magelang, Salatiga, and Pekalongan. This piggy-back system of dependency to a major city has been a formula for small-town economic success and urbanization [4]. In terms of the implementation of regional autonomy policy in Indonesia, this chapter will present a comprehensive examination of land use change and population distribution in Banjar Municipality.

Table 1 Administrative government units in Indonesia, 1985–2015

Administrative level	1985	1990	1995	2000	2005	2010	2013	2014	2015
Province (provinsi)	27	27	27	26	33	33	34	34	34
Regency (kabupaten)	246	241	243	268	349	399	413	416	416
Municipality or city (kota)	55	55	62	73	91	98	98	98	98
District (kecamatan)	3539	3625	3844	4049	5277	6699	6982	7024	7071
Village (desa)	67,534	67,033	65,852	69,050	69,868	77,548	80,714	81,626	81,936

Source Biro Pusat Statistik (i.e., Statistical Bureau), Sixty Years of Indonesian Independence (*Statistik 60 Tahun Indonesia Merdeka*, Statistics to celebrate 60 years of independence of the Republic of Indonesia), and Statistical Yearbook of Indonesia 2016

3   Methodology

3.1   Study Area and Geography

West Java Province, the most populous province in Indonesia, is a key source of development for the country. In the period of almost a decade, the population of West Java Province has increased from 43.2 million in 2010 to 48.7 million in 2019. To best analysis, this population change in conjunction with Indonesia's growing urban corridors, Banjar Municipality, an autonomous city, was chosen as the study area. On February 21, 2002, Banjar Municipality, located in a traffic lane between West Java Province and Central Java Province (and extending all the way through to East Java Province), was inaugurated as an autonomous city by the Minister of Home Affairs. Banjar Municipality, expected to grow as an industrial city, has promising levels of trade, service, and tourism for the eastern region of West Java Province. In 2018, the total population of the Municipality was circa 183,000.

To best piece together Banjar Municipality's prospect for urban change, an overview of its physical geography sheds light on why it is experimentally significant. Its geohydrological conditions are of importance since concern with urban planning will need to consider several types of soil formation and units of rock that make up its lithology. Located at an altitude between 20 and 500 m, alluvium is spread throughout much of the Municipality where flat morphology is present. Frozen rock, andesite in the form of lava, is mainly found within Gunung Sangkur and Babakan Mountain as well as some minor hills along the outskirts (i.e., Pasir Tumpeng, Pasir Leutik, Pasir Kengkol, Pasir Gembok, and Mandalareh–Cadasgintung). The sediment lava, located west of Babakan Mountain, has undulating morphology found throughout the Purwaharja District, making urban development more challenging. Tread formation, which consists of rough green sandstone (i.e., located within the bottom horizon) and sandstone with marlstone thinly insert (i.e., the top horizon), is in the western and the southern parts of the Municipality (i.e., Banjar District and Pataruman District). Soil fertility levels are generally excellent with subtle fine soil texture and alluvial soil types found Municipality-wide, with exception of Langensari District which has red yellow soil which does not affect soil fertility levels. The hydrological condition (i.e., surface water) of Banjar Municipality is sourced from two major rivers (i.e., Citanduy River which spans approximately 14 km and Ciseel River about 15 km). As a result, the territory's watersheds traverse its boundaries collecting rainwater, sediment, and nutrients and draining via the tributaries to the sea or nearby lake. This resource rich, abundantly blessed city has four watersheds (i.e., Citanduy watershed, Cijolang–Citapen watershed, Ciseel–Cikembang–Cimaragas watershed, and Cilisung watershed) which support some of the most fertile and agriculturally productive environments on Earth.

Banjar Municipality, a very unique Indonesian autonomous city, is the only city that administratively is set up with hierarchical sub-divisions (i.e., four districts and 16 villages). Land use change patterns are an elongated pattern which follow the main transecting pathway. Massive levels of development have been implemented since the city was established, creating a successful level of economic growth. In the five years after the city was officially established, the productivity growth rate of the agricultural sector remained relatively high, reaching an average of more than 4% annually, the second major source of economic growth after trade, hospitality (i.e., hotels), and restaurants. The economic growth of Banjar Municipality for the period of 2006–2010 was positive, from 4.71% in 2006 to 4.93% in 2007 and slightly decreasing to 4.82% in 2008 (i.e., mostly due to external factors); however, in 2009, the economic growth skyrocketed to 5.13% and continued to augment to 5.28% in 2010. Banjar Municipality's economic progress, over the last decade, has increasingly gone up with the development of Banjar Waterpark, entertainment services in the Parunglesang area which creates a domino effect of growth to other businesses (e.g., businesses, restaurants, and other hospitality-oriented services). These conditions influenced by urbanization and migration levels, formulate an engaging examination of spatial land use for Indonesia's future development and prospect.

3.2 Data Source and Method

From its founding in 2002, Banjar Municipality has had a number of preliminary land use maps created by the Banjar Municipality Regional Planning and Development Agency in cooperation with the Ministry of Public Works and Public Housing. Data over a ten-year span, from 2006 to 2016, was collected and sourced from the above Agency and Ministry. For 2006, land use data was extrapolated from archived the above Agency and Ministry maps and datasets while, for 2016, changes were assimilated using technology from remote sensing imagery sourced from Google Earth Image Series. Note, other data maps supplied by the US Geological Survey and Global Land Cover Facility were found to be unclear due to the small size of Banjar Municipality (i.e., the whole the territory was covered by green). The use of Google Earth Image Series provided a clear view of buildings, roads, and topology [38, 39]. Malarvizhi et al. [14] state this method as best suited for urban-related applications. Base imagery and population data for this study are illustrated in Figs. 1 and 2.

Using Banjar Municipality Regional Planning and Development Agency's boundary map, the study area was digitized using various land use classes to prepare the land use map. Four land use classes were utilized: urban areas (i.e., built-up land including all buildings and roads), city forests (i.e., forestland in and around the city), agriculture (i.e., comprising of vegetation and agricultural land), and water bodies (i.e., rivers and a natural and artificial lake). An essential advantage of utilizing Google Earth Image Series is that it provides images taken from different periods, which can be useful for urban planning as well as land use change detection. Before digitizing the research, we selected the clearest images from 2016 that could be used in comparing data from 2006.

Next, the digitization process was converted using ArcGIS 10.2 software, an architecture GIS for working with maps and geographic information, from shapefile format (.shp) to Google Earth Image Series compatible format (.kml). Google

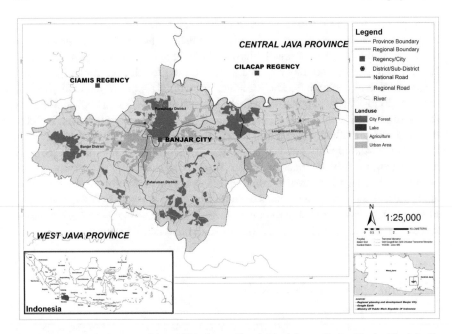

Fig. 1 Analyzed land use condition, 2006. *Source* Google Earth Image Series, Banjar Municipality Regional Planning and Development Agency

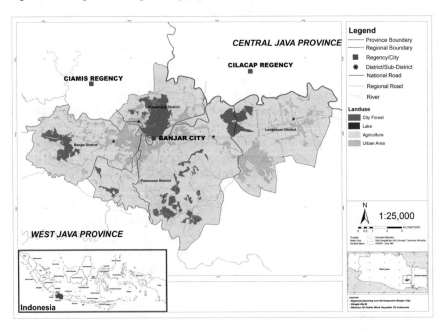

Fig. 2 Analyzed land use condition, 2016. *Source* Google Earth Image Series, Banjar Municipality Regional Planning and Development Agency

Earth software was downloaded and the digitation in.kml format was opened as a boundary area of study. After the image was downloaded into Google Earth software, the image was converted from the geographic coordinate system (i.e., latitude and longitude) to projected coordinate system (i.e., northing and easting) using Universal Transverse Mercator projection in ArcGIS. The projection image was digitized to draw the land use change based on the four land use classes. This process produced a preliminary digitized map of Banjar Municipality's boundary which then involved the cutting of polygons to digitize land use, by inserting 2016 shapefile data for each district (Fig. 3). The final step involved validation. To do so, the local authority was asked to partake and validate with the mapping process of each district. Confirmation of the result compared the digitation process with the official area from the Banjar Municipality Regional Planning and Development Agency. If the result did not correspond with the setting, re-digitization was performed until the

Fig. 3 Process to create the land use change map of Banjar Municipality

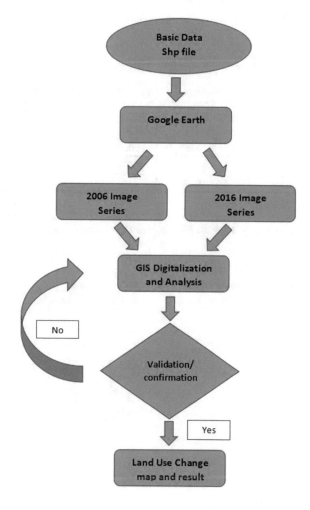

best-case scenario was achieved. Population data collected and analyzed from the Banjar Municipality Regional Planning and Development Agency datasets as well as published census data from the Central Statistics Bureau for the years 2006 and 2016 was used (Table 2). Population data illustrates an increase of the population density subdistrict-wide.

To create the density map, we sub-divided the data into lower-level government administrative units (i.e., *desa* (i.e., rural village) and *kelurahan* (i.e., urban village)). We then divided the process into six classifications according to Sturgess formula (Eq. 1).

$$K = 1 + 3.33 \log n \qquad (1)$$

where 'K' is the classification of the density and 'n' is the *desa* or *kelurahan* number. Formulation was used to classify the boundary in ArcGIS and create the density map (Fig. 4).

Table 2 Population density in Banjar Municipality

District	2006		2016	
	Population	Population density[a]	Population	Population density[a]
Banjar	48,423	1846	58,222	2219
Purwaharja	19,711	1313	24,108	1462
Pataruman	51,348	950	61,173	1205
Langensari	49,430	1480	58,259	1744
Total	168,912		202,362	

[a]Inhabitants per m^2

Source Biro Pusat Statistik Kota Banjar (Statistical Bureau Banjar Municipality), Banjar Municipality Regional Planning and Development Agency

Fig. 4 Process of creating population density map of Banjar Municipality

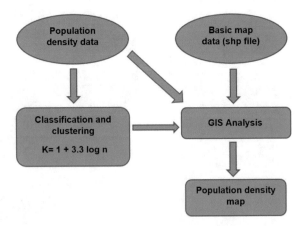

4 Results and Discussion

Since Banjar Municipality became an autonomous city, massive changes in city development—particularly in urban areas—have experienced an increase in population density. This is confirmed by the Banjar Municipality Regional Planning and Development Agency and the Ministry of Public Works and Public Housing. Population density concerns have come at the expense of agricultural area (Fig. 5). In consequence, between 2006 and 2016 that the overall urban area of the Municipality has increased from 13.49 to 15.41% while agricultural area decreased from 71.22 to 69.87%. Interestingly, the city forested area has had only minimal impact while the water bodies have slightly augmented due to extensions in construction of the artificial lake—facilitated by the local government.

Urban sprawl in Banjar Municipality is concentrated in two districts, namely Banjar and Langensari. This is understandable since the infrastructure development policy, from the local authority, places Banjar District as the city center and Langensari District as the key industrial area. With exception to the increase in water body area, most of the changes hampered agricultural lands in both districts. Despite this, most of the urban change occurring within these two districts, historically from its onset, the autonomous city's urban growth was evenly spread out. However, urbanization trends still indicate a positive city plan and direction for its future. At present,

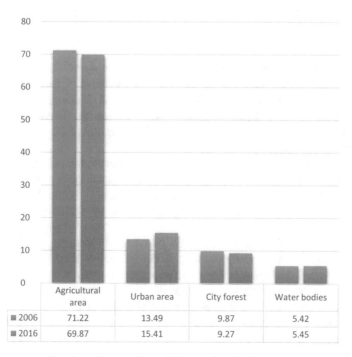

Fig. 5 Percentage of land use change in Banjar Municipality, 2006 and 2016

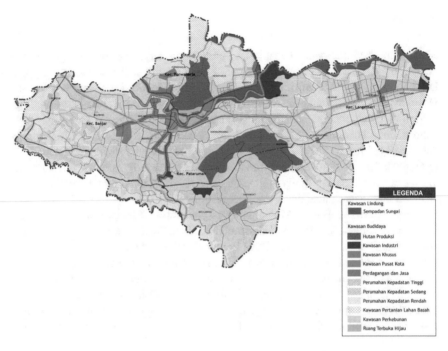

Fig. 6 Banjar Municipality city plan, 2008. *Source* Banjar Municipality Planning and Development Agency

city-wide research indicates consistent and sound development strategies, something urban planners and managers alike have seen as a positive sign since its transformation as an autonomous city. Self-determination-based planning, in this sense, can be seen as an encouraging developmental direction for investment and economic productivity. Controlled urbanized planning will be central to the Municipality and its future prospect (Fig. 6).

According to the urban area map from 2016, concentric zoning [40, 41] of the city does not appear to be present in Banjar Municipality. Since the Municipality's transformation into an autonomy city, the urban sprawl pattern has not been drastically changed (Fig. 7). Correspondingly, the population density map illustrates that urban area change occurred in the same location as the increase of population density; however, it did not transpire in the highest populated district (Fig. 8). These results indicate that urban area mostly occurred in the districts of Banjar and Langensari. One of the essential findings is that when population density is classified at a lower administrative government level, these administrative governments can focalize their attention of micro-managing their areas and, to some degree, complete with neighboring villages for resources and productivity. This is evident from the 2016 findings where population density increased in two urban villages, namely Mekarsari and Hegarsari, which are located in the two infrastructure-favored districts of Banjar and Pataruman, not in the highest populated.

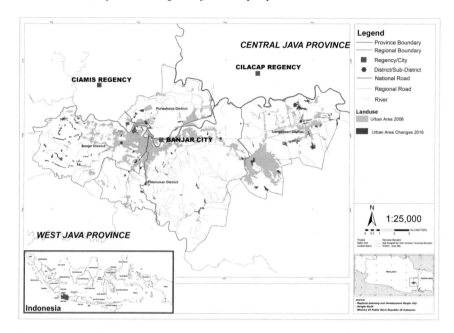

Fig. 7 Urban area change in Banjar Municipality, 2016

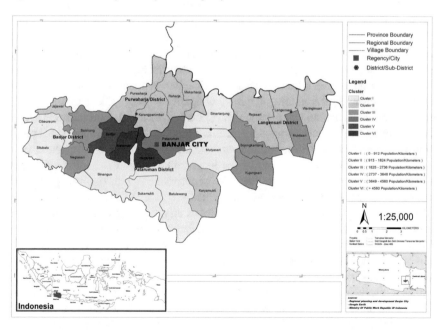

Fig. 8 Population density in Banjar Municipality, 2016

5 Conclusion

A positive correlation between population and land use change such as in the district of Somba Opu [42] cannot be generalized for Banjar Municipality. According to the results of the study, the land use change map indicates significant increase in change occurring along the lower parts of the city, namely Banjar District and Langensari District, while Pataruman District which is considered to have the highest population will not change as much or quickly. This situation is in line with previous research conducted throughout Jilin Province, China [43].

An imbalanced economic spread in living and habitation standards has highly contributed to the significant gap between the increase in population and availability to new urban residents. Action should be taken to prioritize sound urban development and proper management and plan an environmentally friendly cityscape. Moreover, regional development and decentralized institutions that encourage local public authorities to be more active in developing their environment with local-based knowledge should be pursued. This practice has repetitively been confirmed as positive and measurable. As such, impact on improving governance in a developing country, as seen in Banjar Municipality, requires playing close attention to detail as resources are often limited or restricted.

Empirical evidence that the impact of globalization on the Municipality's development constricts agriculture to the territory's outskirts. This is common throughout much of the world as more people are continually moving from rural to urban environments [44–46]. In an Indonesian sense, this condition is unique in that change is locally gauged by the Banjar Municipality which does not have overly-constricting external authority. As an autonomous city, its administrative sovereignty likens its development pattern as a hub for trade, service, and tourism in eastern West Java Province. However, the nature of the Municipality's geographical makeup as an urban corridor and bridge-city between West Java Province and Central Java Province defines much of the linear transitory results found. Attention to land use change impacts are critical and urbanization unavoidable. Unsound, environmental concerns should be closely monitored by the local authority which should continue to pay close attention to water resources, their conservation with rigorous protection measure, and set watershed management guidelines that conserve the new artificial lake's water supply. As such, as an autonomous city, not dependent on the central government, local know-how can be advocated for community-driven best practices. To achieve comprehensive development, further research should be conducted in comparably similar areas. Research should look at patterns of land use population density and expand the array of land use classes to include: socioeconomics, gross domestic product, social welfare and health, community participation, and education. These other elements enlarge the dimensional status of the research and would be beneficial to urban knowledge-base development and transformation thinking.

Acknowledgements The authors would like to acknowledge support from the China Scholarship Council, National Natural Science Foundation of China (Grant No: 41471103), West Java Province Government, and Banjar Municipality Government.

References

1. Huilei L, Jian P, Yanxu L, Yi'na H (2017) Urbanization impact on landscape patterns in Beijing City, China: a spatial heterogeneity perspective. Ecol Indic 82:50–60. https://doi.org/10.1016/j.ecolind.2017.06.032
2. Demographia (2018) Demographia world urban areas & population projections
3. World Bank (2016) Indonesia's urban story
4. Prawatya NA (2013) Spatial development in small cities in Middle Java Province. J Wil Dan Lingkung 1:17–32
5. Ardiwijaya VS, Sumardi TP, Suganda E, Temenggung YA (2015) Rejuvenating idle land to sustainable urban form: case study of Bandung metropolitan area, Indonesia. Procedia Environ Sci 28:176–184. https://doi.org/10.1016/j.proenv.2015.07.024
6. Ruan X, Qiu F, Dyck M (2016) The effects of environmental and socioeconomic factors on land-use changes: a study of Alberta, Canada. Environ Monit Assess 188:466. https://doi.org/10.1007/s10661-016-5450-9
7. Long H, Heilig GK, Li X, Zhang M (2007) Socio-economic development and land-use change: analysis of rural housing land transition in the transect of the Yangtse River, China. Land Use Policy 24:141–153. https://doi.org/10.1016/j.landusepol.2005.11.003
8. Fang Y-G, Shi K-J, Niu C-C (2016) A comparison of the means and ends of rural construction land consolidation: case studies of villagers' attitudes and behaviours in Changchun City, Jilin Province, China. J Rural Stud 47:459–473. https://doi.org/10.1016/j.jrurstud.2016.04.007
9. Li H, Yuan Y, Zhang X et al (2019) Evolution and transformation mechanism of the spatial structure of rural settlements from the perspective of long-term economic and social change: a case study of the Sunan region, China. J Rural Stud 1. https://doi.org/10.1016/j.jrurstud.2019.03.005
10. Yuan F, Wei YD, Chen W (2014) Economic transition, industrial location and corporate networks: remaking the Sunan Model in Wuxi City, China. Habitat Int 42:58–68. https://doi.org/10.1016/j.habitatint.2013.10.008
11. Potere D, Schneider A, Angel S, Civco DL (2009) Mapping urban areas on a global scale: which of the eight maps now available is more accurate? Int J Remote Sens 30:6531–6558. https://doi.org/10.1080/01431160903121134
12. Hadi F, Thapa RB, Helmi M, Hazarika MK (2016) Urban growth and land use: land cover modeling in Semarang, Central Java, Indonesia
13. Azam M (2019) Accounting for growing urban-rural welfare gaps in India. World Dev 122:410–432. https://doi.org/10.1016/j.worlddev.2019.06.004
14. Malarvizhi K, Kumar SV, Porchelvan P (2016) Use of high resolution Google Earth satellite imagery in landuse map preparation for urban related applications. Procedia Technol 24:1835–1842. https://doi.org/10.1016/j.protcy.2016.05.231
15. Sidhu N, Pebesma E, Câmara G (2018) Using Google Earth engine to detect land cover change: Singapore as a use case. Eur J Remote Sens 51:486–500. https://doi.org/10.1080/22797254.2018.1451782
16. Hu Q, Wu W, Xia T et al (2013) Exploring the use of Google Earth imagery and object-based methods in land use cover mapping. Remote Sens 5:6026–6042. https://doi.org/10.3390/rs5116026
17. Widyasamratri H, Aswad A (2017) A preliminary study: an agent-based spatial simulation of human-coastal environment interaction. In: The third international conference on coastal and delta areas, pp 593–601
18. Handayani W, Rudiarto I (2011) Dinamika Persebaran Penduduk Jawa Tengah: Perumusan Kebijakan Perwilayahan Dengan Metode Kernel Density. Diponegoro University
19. Held D, McGrew A, Goldblatt D, Perraton J (1999) Introduction in global transformations. In: Global transformations: politics, economics and culture. Stanford University Press, Stanford, pp 1–31
20. Panennungi MA (2013) Cost and benefit of globalization: lesson learned from Indonesian history

21. Yuniarto PR (2014) Globalization problem in Indonesia: between interest, policy and challenge. J Kaji Wil 5(29)
22. Kis-Katos K, Sparrow R (2011) Child labor and trade liberalization in Indonesia. J Hum Resour 46:722–749. https://doi.org/10.1353/jhr.2011.0008
23. Keohane RO, Nye JSJ (1998) The Oxford dictionary of quotations. Harvard University
24. Vujakovic P (2010) How to measure globalization? A new globalization index (NGI). Atl Econ J 38:237. https://doi.org/10.1007/s11293-010-9217-3
25. Rifai M (2017) Regional autonomy and globalization: a development model study in Karawang Regency. J Polit Indones 2:15–28
26. Stasavage D (2014) Was Weber right? The role of urban autonomy in Europe's rise. Am Polit Sci Rev 108:337–354. https://doi.org/10.1017/S0003055414000173
27. Paierele O (2011) Globalization impact on the local and regional development policies: Republic of Moldova case
28. Myerson RB (2015) Democratic decentralization and economic development. In: The Oxford handbook of Africa and economics, pp 1–19. https://doi.org/10.1093/oxfordhb/9780199687114.013.18
29. Riadi B, Barus B, Widiatmaka et al (2018) Identification of flood area in the coastal region using remote sensing in Karawang Regency, West Java. In: IOP conference series: earth and environmental science
30. Darmawan R (2008) The practices of decentralization in Indonesia and its implication on local competitiveness. University of Twente
31. Rudy, Hasyimzum Y, Heryandi, Khoiriah S (2017) 18 years of decentralization experiment in Indonesia: institutional and democratic evaluation. J Polit Law 10:132–139. https://doi.org/10.5539/jpl.v10n5p132
32. Mauro L, Pigliaru F, Carmeci G (2018) Decentralization and growth: do informal institutions and rule of law matter? J Policy Model 40:873–902. https://doi.org/10.1016/j.jpolmod.2018.05.003
33. Ishak AF, Utomo TWW (2014) Decentralization policy and regional autonomy implementation in East Kalimantan. J Borneo Adm 6a:1–17. https://doi.org/10.24258/jba.v6i2.57
34. Miller MA (2009) Rebellion and reform in Indonesia: Jakarta's security and autonomy policies in Aceh. J Curr Southeast Asian Aff 28(4):145–151. https://doi.org/10.1159/000118017
35. Miller MA (2013) Decentralizing Indonesian city spaces as new "centers". Int J Urban Reg Res 37:834–848. https://doi.org/10.1111/j.1468-2427.2013.01209.x
36. Phelps NA, Bunnell T, Miller MA (2011) Post-disaster economic development in Aceh: neoliberalization and other economic-geographical imaginaries. Geoforum 42:418–426. https://doi.org/10.1016/j.geoforum.2011.02.006
37. Rodríguez-Pose A, Ezcurra R (2010) Does decentralization matter for regional disparities? A cross-country analysis. J Econ Geogr 10:619–644. https://doi.org/10.1093/jeg/lbp049
38. Ludwig A, Meyer H, Nauss T (2016) Automatic classification of Google Earth images for a larger scale monitoring of bush encroachment in South Africa. Int J Appl Earth Obs Geoinf 50:89–94. https://doi.org/10.1016/j.jag.2016.03.003
39. Qi F, Zhai JZ, Dang G (2016) Building height estimation using Google Earth. Energy Build 118:123–132. https://doi.org/10.1016/j.enbuild.2016.02.044
40. Burgess EW (2008) The growth of the city: an introduction to a research project. In: Urban ecology. Springer US, Boston, MA, pp 71–78
41. Shertzer A, Twinam T, Walsh RP (2018) Zoning and the economic geography of cities. J Urban Econ 105:20–39. https://doi.org/10.1016/j.jue.2018.01.006
42. Syaifuddin, Hamire A, Dahlan (2013) The relationship between and land use change in Somba Opu District, Gowa Regency. J Agrisistem 9:169–179
43. Li F, Zhang S, Bu K et al (2015) The relationships between land use change and demographic dynamics in western Jilin Province. J Geogr Sci 25:617–636. https://doi.org/10.1007/s11442-015-1191-x
44. Russo A, Cirella G (2018) Modern compact cities: how much greenery do we need? Int J Environ Res Public Health 15:2180. https://doi.org/10.3390/ijerph15102180

45. Brøgger D (2019) Urban diaspora space: rural-urban migration and the production of unequal urban spaces. Geoforum 102:97–105. https://doi.org/10.1016/j.geoforum.2019.04.003
46. Chen T, Lang W, Chan E, Philipp CH (2018) Lhasa: urbanising China in the frontier regions. Cities 74:343–353. https://doi.org/10.1016/j.cities.2017.12.009

Glossary

Anthropocentrism Refers to a human-centered, or "anthropocentric," point of view. In philosophy, anthropocentrism can refer to the point of view that humans are the only, or primary, holders of moral standing. Anthropocentric value systems thus see nature in terms of its value to humans; while such a view might be seen most clearly in advocacy for the sustainable use of natural resources even arguments that advocate for the preservation of nature on the grounds that pure nature enhances the human spirit must also be seen as anthropocentric [1].

Biodiversity Biological diversity (i.e., biodiversity) is the variability among living organisms from all sources including, inter alia, terrestrial, marine and other aquatic ecosystems, and the ecological complexes of which they are part; this includes diversity within species, between species and of ecosystems [2].

Carrying capacity The maximum population that can be supported by a given set of resources [3].

Climate change Any change in the state of the climate that can be identified by changes in the mean and/or variability of its properties, and that persists for an extended period, typically decades or longer [4].

Cohesion policy The European Union's main investment policy. It targets all regions and cities in the European Union in order to support job creation, business competitiveness, economic growth, sustainable development, and improve citizens' quality of life [5].

Conservation Refers to any form of environmental protection, including preservation [3]. The careful management and use of natural resources, the achievement of significant social benefits from them, and the preservation of the natural environment [6].

Consumption Fixed on a series of social-cum-cultural issues in relation to the way that commodities and their meaning have become intertwined. Specifically, the extent to which a common global capitalist culture has been created by ever-increasing circulation of commodities and commodity meanings around the world. The salability and commodification of everything [7].

© Springer Nature Singapore Pte Ltd. 2020
G. T. Cirella (ed.), *Sustainable Human–Nature Relations*,
Advances in 21st Century Human Settlements,
https://doi.org/10.1007/978-981-15-3049-4

Darwinism The theory of the evolutionary a mechanism propounded by Charles Darwin as an explanation of organic change. It denotes Darwin's specific view that evolution is driven mainly by natural selection [8].

Decentralization The transfer of authority and responsibility for public functions from the central government to intermediate and local governments or quasi-independent government organizations or the private sector [9].

Development The economic, social, and institutional evolution of national states [6]. Can be interpreted to mean the process of becoming larger, more mature, and better organized—often in terms of economic criteria [3].

Doing business Advocates for both regulatory quality and efficiency with effective rules in place that are easy to follow and understand. The term refers to being able to realize economic gains, reduce corruption, and encourage SMEs to flourish. It also implies unnecessary red tape should be eliminated. Specific safeguards must also be put in place to ensure high-quality business regulatory processes [10].

Ecocentrism Positions the ecosphere—comprising all Earth's ecosystems, atmosphere, water, and land—as the matrix which birthed all life and as life's sole source of sustenance. It is a worldview that recognizes intrinsic value in ecosystems and the biological and physical elements that they comprise, as well as in the ecological processes that spatially and temporally connect them. Ecocentrism contrasts sharply with anthropocentrism [11].

Ecology The study of relationships between organisms and their environments [3].

Ecosystem services The direct and indirect contributions of ecosystems to human well-being. There are four types of ecosystem services: provisioning, regulating, supporting, and cultural [12].

Entrepreneurialism Defined as starting new businesses, or getting involved with new ventures or ideas [13].

Focus group discussion A qualitative research method in the social sciences with a particular emphasis and application in the developmental program evaluation sphere [14].

Forced displacement Forced displacement is a development challenge, not only a humanitarian concern. The involuntary or coerced movement of a person or people away from their home or home region, resulting from causes including natural disasters, violence, and persecution [15].

Gaia hypothesis All living organisms and their inorganic surroundings evolved together as a single, self-regulating system that has kept the planet habitable for life—despite threats such as a brightening Sun, volcanoes, and meteorite strikes. Introduced in the 1970s, by James Lovelock and Lynn Margulis, the concept is based on Earth itself being a living system or organism [16, 17].

Global North Represent the economically developed societies of Europe, North America, Australia, Israel, South Africa, Japan, South Korea, among others. Global North countries are wealthy, technologically advanced, politically stable, and aging as their societies tend toward zero population growth [18].

Global South Represent the economically backward countries of Africa, Central and South America, the Indian subcontinent, the Middle East, among others.

The Global South countries are agrarian based, dependent economically and politically on the Global North the Global North has continued to dominate and direct the Global South in international trade and politics [18].

Green infrastructure A strategically planned network of high-quality natural and semi-natural areas with other environmental features, which is designed and managed to deliver a wide range of ecosystem services and protect biodiversity in both rural and urban settings [19].

Habitation The state of living somewhere. When an area has no human habitation, it means that no people live there [20].

Human settlement A form of human habitation which ranges from a single dwelling to large urban center. It is a process of opening up and settling of a previously uninhabited area by people. The study of human settlement is basic to human geography as it examines the forms of settlement in any particular region reflective of the human relationship with the environment [21].

Inequality Occurs when resources in a given society are distributed unevenly, typically through norms of allocation, that engender specific patterns along lines of socially defined categories of persons. It is the differentiation preference of access of social goods in the society brought about by power, religion, kinship, prestige, race, ethnicity, gender, age, sexual orientation, and class. Often referred to as social or economic inequality [22, 23].

Integration The two-way process of mutual adaptation between migrants and the societies in which they live, whereby migrants are incorporated into the social, economic, cultural, and political life of the receiving community. It entails a set of joint responsibilities for migrants and communities, and incorporates other related notions such as social inclusion and social cohesion [24].

Irregular migration Movement of persons that takes place outside the laws, regulations, or international agreements governing the entry into or exit from the state of origin, transit, or destination [24].

Land speculation In commerce, the act or practice of buying land in expectation of a rise of price and of selling it at an advance, as distinguished from a regular trade, in which the profit expected is the difference between the wholesale and retail price [25].

Livelihood Can be defined as the methods and means of making a living. The notion includes access to resources such as land or property, crops, food, knowledge, finance, social relationships, and their interrelated connection with the political, economic, and sociocultural characteristics of an individual community. A livelihood consists of capabilities, assets, and activities that are required for living [26]. A sustainable livelihood interlinks the definitions of resilience, sustainability, and offsets reducing vulnerabilities of the community, including reducing poverty levels, building capacities, and coping mechanisms, and focusing on community resilience [27].

Marginalization A form of acute and persistent disadvantage rooted in underlying social inequalities [28].

Methodology Is a theory of producing knowledge through research and provides a rationale for the way a researcher proceeds. It is the philosophical underpinning of a given research practice. Methodology is the discussion of the theory upon which research is based and indicates how the experimentation is performed [29].

Nature-based solutions Actions to protect, sustainably manage, and restore natural or modified ecosystems, that address societal challenges effectively and adaptively, simultaneously providing human well-being and biodiversity benefits [30, 31].

Opportunity cost Represent the benefits an individual, investor, or business misses out on when choosing one alternative over another [32]. The two definitions of opportunity cost differ in what is forgone. For the "quantity" type, it is the highest-valued alternative (i.e., the physical thing or things that otherwise would have been chosen). For the "value" type, it is the value of the highest-valued alternative (i.e., the value of the physical thing or things that otherwise would have been chosen) [33].

Protected area A clearly defined geographical space, recognized, dedicated, and managed, through legal or other effective means, to achieve the long-term conservation of nature with associated ecosystem services and cultural values [34].

Resilience The process of adapting well in the face of adversity, trauma, tragedy, threats, or even significant sources of risk. The act of bouncing back or resisting under pressure. Resilience theory is a set of ideas that discuss the impact of challenging events on individuals, communities, or any level of society and how well they have adapted to a traumatic experience [35, 36].

Resource A concept used to denote sources of human satisfaction, wealth, or strength. Labor, entrepreneurial skills, investment funds, fixed capital assets, technology, knowledge, social stability, and cultural and physical attributes may be referred to as the resources for a country. In a resource management context, natural resources, which are substances, organisms, and properties of the physical environment, are valued for their perceived ability to satisfy human needs and wants [7].

Social cohesion The willingness of members of a society to cooperate with each other in order to survive and prosper [37].

Social sustainability Adaption and integration of precautionary social principles and considerations into decision-making processes. Social sustainability provides a basic architecture for a community to develop processes and structures which not only meet the needs of its current members but also support the ability of future generations to maintain a healthy community [38].

Sustainability Development that meets the needs of the present without compromising the ability of future generations to meet their own needs [39].

Traditional value The moral and ethical principles traditionally upheld and passed on to promote the sound functioning of and fabric of society [40].

Urban green space Urban space covered by vegetation of any kind, including smaller green space features (e.g., street trees and roadside vegetation), green spaces not available for public access or recreational use (e.g., green roofs and facades, or green space on private grounds), and larger green spaces that provide various social and recreational functions (e.g., parks, playgrounds, or greenways) [41].

Urban heat island Phenomenon in which cities and towns are much warmer than surrounding rural areas, particularly at night. It is primarily generated as a result of the physical properties of urban materials, their structure and to a lesser extent their use, e.g., through anthropogenic heat emissions [25].

Urban sprawl Refers to urbanization and the migration of a population from populated towns and cities to low density residential development over more and more rural land. The end result is the spreading of a city and its suburbs over such land. Urban sprawl can be low-density residential and commercial development moved to undeveloped land [42]. Largely unplanned expansion of an urban area, typically discontinuous, leaving rural enclaves [3].

Urban sustainability Based on positive interactions among three different urban substrates: physical, social, and economic. Urban sustainability predominately focuses on the city in which social well-being coexists with economic development and environmental quality [43].

Utilitarianism A moral theory which proposes the maximization of human well-being (i.e., utility or welfare) as the goal of life [7].

Vulnerability Within a migration context, it is the limited capacity to avoid, resist, cope with, or recover from harm. This limited capacity is the result of the unique interaction of individual, household, community, and structural characteristics and conditions [24].

Well-being The degree to which the needs and wants of a society are satisfied [3]. An individual's perception of their position in life in the context of the culture and value systems in which they live and in relation to their goals, expectations, standards, and concerns. It is a broad ranging concept affected in a complex way by the person's physical health, psychological state, personal beliefs, social relationships, and their relationship to salient features of their environment [44].

References

1. Padwe J (2016) Anthropocentrism. In: Oxford bibliogr. https://brill.com/view/title/18825. Accessed 30 Oct 2019
2. United Nations Convention on Biological Diversity (1992) Convention on biological diversity. https://www.cbd.int/doc/legal/cbd-en.pdf. Accessed 30 Oct 2019
3. Norton W (1998) Human geography. Oxford University Press
4. IPCC (2007) Climate change 2007: The physical science basis. https://www.ipcc.ch/report/ar4/wg1/. Accessed 30 Oct 2019
5. European Commission (2014) An introduction to EU Cohesion Policy 2014–2020. European Commission, London
6. De Blij HJ, Murphy AB (2003) Human geography: culture, society, and space, 7th edn. Wiley, Hoboken
7. Johnston R, Gregory D, Pratt G, Watts M (2006) The dictionary of human geography, 4th ed. Gregory, Pratt Johnston: 9781405136044: Amazon.com: Books. Wiley, Malden
8. Encyclopaedia Britannica (2019) Darwinism. In: Encyclopedia Britannica. https://www.britannica.com/science/Darwinism. Accessed 30 Oct 2019
9. World Bank (2001) What, why, and where. In: World bank. http://www1.worldbank.org/publicsector/decentralization/what.htm. Accessed 30 Oct 2019
10. World Bank (2019) Doing business 2019: training for reform, 16th edn. World Bank, Washington
11. Gray J, Whyte I, Curry P (2018) Ecocentrism: what it means and what it implies. Ecol Citiz 1:130–131
12. TEEB (2019) Ecosystem services. In: Economics of ecosystems and biodiversity. http://www.teebweb.org/resources/ecosystem-services/. Accessed 30 Oct 2019
13. Your Dictionary (2018) Entrepreneurialism. In: LoveToKnow Corp. https://www.yourdictionary.com/entrepreneurialism. Accessed 30 Oct 2019
14. Humans of Data (2017) How to conduct a successful focus group discussion. In: Humans data. https://humansofdata.atlan.com/2017/09/conduct-successful-focus-group-discussion/. Accessed 30 Oct 2019
15. Martin SF (2018) Forced migration and refugee policy. In: Hugo G, Abbasi-Shavazi MJ, Kraly EP (eds) Demography of refugee and forced migration. Springer International Publishing, Cham, pp 271–303
16. Lenton TM, Latour B (2018) Gaia 2.0. Science (80–) 361:1066–1068. https://doi.org/10.1126/science.aau0427
17. Onori L, Visconti G (2012) The GAIA theory: from Lovelock to Margulis. From a homeostatic to a cognitive autopoietic worldview. Rend Lincei 23:375–386. https://doi.org/10.1007/s12210-012-0187-z

18. Odeh L (2012) A comparative analysis of global north and global south economies. Int J Curr Res Humanit 12:67–82
19. Directorate-General for Environment (2013) Building a green infrastructure for Europe. In: Eur. Comm. https://op.europa.eu/en/publication-detail/-/publication/738d80bb-7d10-47bc-b131-ba8110e7c2d6/language-en. Accessed 30 Oct 2019
20. Vocabulary.com (2019) Habitation. In: Vocabulary.com. https://www.vocabulary.com/dictionary/habitation. Accessed 30 Oct 2019
21. Josh J (2015) Human settlement. https://www.jagranjosh.com/general-knowledge/human-settlement-1448446636-1. Accessed 30 Oct 2019
22. Wikipedia (2019) Social inequality. https://en.wikipedia.org/wiki/Social_inequality. Accessed 30 Oct 2019
23. Neckerman KM, Torche F (2007) Inequality: causes and consequences. Annu Rev Sociol 33:335–357. https://doi.org/10.1146/annurev.soc.33.040406.131755
24. IOM (2019) Key migration terms. In: International organization for migration https://www.iom.int/key-migration-terms. Accessed 30 Oct 2019
25. Lindley SJ, Cook PA, Dennis M, Gilchrist A (2019) Biodiversity, physical health and climate change: a synthesis of recent evidence. Biodiversity and health in the face of climate change. Springer International Publishing, Cham, pp 17–46
26. Islam T, Ryan J, Islam T, Ryan J (2016) Mitigation in the private sector. Hazard Mitig Emerg Manag 101–124. https://doi.org/10.1016/B978-0-12-420134-7.00004-7
27. UNESCAP (2008) Greening growth in Asia and the Pacific. United Nations Economic and Social Commission for Asia and the Pacific, Bangkok
28. UNESCO (2010) Reaching the marginalized. Oxford University Press, Oxford
29. Zeegers M, Barron D, Zeegers M, Barron D (2015) Milestone 5: Methodology. Milestone moments get your PhD Qual Res 61–74. https://doi.org/10.1016/B978-0-08-100231-5.00005-5
30. IUCN (2019) Nature-based solutions. In: International Union for conservation of nature. https://www.iucn.org/commissions/commission-ecosystem-management/our-work/nature-based-solutions. Accessed 30 Oct 2019
31. Cohen-Shacham E, Walters G, Janzen C, Maginnis S (2016) Nature-based solutions to address societal challenges. IUCN, Gland
32. Kenton W (2019) Opportunity cost. In: Investopedia. https://www.investopedia.com/terms/o/opportunitycost.asp. Accessed 30 Oct 2019
33. Parkin M (2016) Opportunity cost: a reply. J Econ Educ 47:35–39. https://doi.org/10.1080/00220485.2015.1106366
34. IUCN (2008) International Union for conservation of nature website. http://www.iucn.org/. Accessed 13 Dec 2017
35. Richardson GE, Neiger BL, Jensen S, Kumpfer KL (1990) The resiliency model. Health Educ 21:33–39. https://doi.org/10.1080/00970050.1990.10614589
36. Greene RR, Galambos C, Lee Y (2004) Resilience theory. J Hum Behav Soc Environ 8:75–91. https://doi.org/10.1300/J137v08n04_05
37. Stanley D (2003) What do we know about social cohesion: the research perspective of the Federal Government's social cohesion research network. Can J Sociol 28:5. https://doi.org/10.2307/3341872
38. EMG (2019) Environmental and social sustainability. In: United Nations Environ. Manag. Gr. https://unemg.org/our-work/internal-sustainability/environmental-and-social-sustainability/. Accessed 30 Oct 2019
39. Brundtland GH (1987) Our common future: the world commission on environment and development. Oxford University Press, Oxford
40. Galland O, Lemel Y (2008) Tradition versus modernity: the continuing dichotomy of values in European Society. Rev Française Sociol 49:153–186
41. Hunter RF, Cleary A, Braubach M (2019) Environmental, health and equity effects of urban green space interventions. Biodiversity and health in the face of climate change. Springer International Publishing, Cham, pp 381–409

42. Conserve Energy Future (2019) Causes and effects of urban sprawl. In: Conserve energy future https://www.conserve-energy-future.com/causes-and-effects-of-urban-sprawl.php. Accessed 30 Oct 2019
43. Diappi L, Bolchi P, Franzini L (1999) Urban sustainability: complex interactions and the measurement of risk. Cybergeo. https://doi.org/10.4000/cybergeo.1240
44. WHO (2014) WHOQOL: measuring quality of life. In: World Health Organization https://www.who.int/healthinfo/survey/whoqol-qualityoflife/en/. Accessed 30 Oct 2019

Index

Printed in the United States
by Baker & Taylor Publisher Services